ADVANCES IN
X-RAY CONTRAST
Collected Papers

ADVANCES IN X-RAY CONTRAST
Collected Papers

Edited by

P. Dawson
Department of Imaging
Royal Postgraduate Medical School
Hammersmith Hospital
London, UK

and W. Clauss
Schering AG
Clinical Development
Monomeric X-Ray Contrast Media
Berlin, Germany

SPRINGER SCIENCE+BUSINESS MEDIA, B.V.

Library of Congress Cataloging-in-Publication Data is available.

ISBN 978-0-7923-8741-1   ISBN 978-94-011-3959-5 (eBook)
DOI 10.1007/978-94-011-3959-5

Printed on acid-free paper

Contents

Preface
Peter Dawson   1

On the nephrotoxic potential of the iodinated intravascular contrast agents
Peter Dawson   3

Nephrotoxicity related to X-ray contrast media
Knut Joachim Berg and Jarl Å Jacobsen   11

The role of contrast agents in thromboembolic phenomena in clinical angiography
Peter Dawson   20

Delayed reactions to intravenous injections of urographic contrast media
Peter Davies   29

Contrast media use in paediatrics
Paul Babyn   34

Cardiac use and effects of contrast agents
Michael Bettmann   46

Contrast agents in interventional radiology
Peter Dawson   52

Contrast enhancement in computed tomography of the liver, pancreas and spleen
Andreas Adam   57

Spiral computed tomography – a short overview
Mathias Langer   63

Electron Beam Computed Tomography (EBCT)
R Knapp, I Bangerl, D zur Nedden   67

Iodinated contrast agents in neuroradiology
Ronald J Zagoria   81

Development of intravascular contrast agents: the first 100 years
Ronald G Grainger   89

Risk factors, prophylaxis and therapy of X-ray contrast media reactions
William H Bush, Jr   97

Contrast medium administration in spiral computerised tomography: an overview of a consensus
meeting in radiodiagnosis
S Feuerbach   107

Index   111

Introduction

For all that new non-X-ray technologies such as MR and ultrasound and its various manifestations have made an enormous impact in recent years on the practice of medical imaging, the use of X-rays and X-ray contrast-enhancing agents has retained an important position at the heart of the process. Indeed, with its frequent requirements for high total dose regimes, CT has increased the the use of contrast agents. Even helical/spiral CT which, it was initially argued, should reduce contrast as well as radiation loads may actually require just as much or more of both cause of the potential it offers for multi-phase scanning.

Iodinated intravascular X-ray contrast agents, especially the more recently developed non-ionic agents, continue therefore to play a pivotal role in clinical imaging.

These succinct and authoritative articles, originally appearing in the *Advances in X-ray Contrast* journal series, range sufficiently widely for their compilation in this volume to be considered a mini-textbook on the water-soluble iodinated X-ray contrast agents and their applications. Each is written by an acknowledged and experienced expert in the field. They usefully cover the developmental history of the agents; defined risk factors, approaches to prophylaxis and, ultimately, of the treatment of adverse reactions; the interesting subject of supposed delayed reactions to contrast agents; the important organ-specific toxicities, cardiac toxicity, neurotoxicity and nephrotoxicity and high-dose toxicity as encountered in complex procedures; the sometimes special circumstances and occasional extreme conditions to which contrast agents may be exposed in Interventional Radiology; the special in several ways case of paediatric radiology; the controversial subject of thrombo-embolic phenomena in clinical angiography; and the precise role of contrast agents. As regards the practicalities of contrast administration regimes and imaging protocols it is really only in the area of CT that there is debate and controversy, and articles are included which cover CT of the liver, spleen and pancreas, and protocols for the new spiral/helical technology and even for the much less widely available electron-beam CT technology visualization, and pulmonary embolus diagnosis, and protocols for contrast administration with this technology are also discussed.

This book should provide either a good general introduction to the subject of X-ray contrast agents for the radiology trainee, or a quick and readable refresher course for the established radiologist or for clinicians such as cardiologists, nephrologists and neurologists with an interest in contrast agents.

Dr Peter Dawson

P. Dawson and W. Clauss, (eds.), Advances in X-Ray Contrast: Collected Papers. 3–10
© *1998 Kluwer Academic Publishers.*

On the nephrotic potential of the iodinated intravascular agents

P Dawson, PhD, MRCP, FRCR
Royal Postgraduate Medical School, Hammersmith Hospital, Du Cane Road, London, W12 ONN, UK

A little more than 60 years ago, a young American, Moses Swick, working in the department of the urologist von Lichtenberg in Berlin, used a mono-iodinated pyridine compound to perform an intravenous urogram [1]. Since then, there have been several generations of iodinated intravascular contrast agents, the latest being the low osmolality ionic and non-ionic types [2,3]. In spite of the development of new imaging techniques, the scale of use and, relevantly, the doses used in many examinations have tended to increase.

Contrast agent-associated nephrotoxicity, broadly defined as an acute impairment of renal function associated with the administration of radiological contrast media, other factors having been eliminated, was first recognized in the 1930s after the widespread adoption of urographic and angiographic contrast procedures using the (then) new agents. In the earliest days, the general systemic adverse reactions (pain, vomiting, etc.) to these agents dominated the clinical picture and contrast medium-induced nephropathy received little attention until about 1970. Even in the period from the mid 1950s to the mid 1970s, the heyday of the ionic, high osmolar, ratio 1.5 contrast media, the number of reported contrast-medium associated renal events was apparently in single figures.

However, the general toxicity profile of these agents was considerably better than that of previously used materials and this obviously led to a more liberal use of contrast media, including the use of higher doses. This trend has continued further with the more recent introduction of 'low osmolality' agents [2,3]. In this context, it is worth noting that over the last half century there have been several changes in the perception of this whole subject. In the 1950s it was widely believed that urography was relatively contraindicated in the face of renal impairment because the contrast agents were likely to engender a further deterioration in renal function and, in the case of urography, unlikely to yield any useful information. By the early 1960s came the realization that, with a sufficiently high dose of contrast agent in renal failure, it was usually possible to produce a nephrogram and to exclude obstruction, important information capable of changing patient management. Macewan et al [4] and Schwartz et al [5] demonstrated this in children and adults, and a number of others reported similar findings [6–9]. These latter authors reported no evidence of reduced renal function and by the mid 1970s high-dose urography was accepted as useful and safe, provided that the patient was hydrated. However, this early evidence for the safety of contrast administration as regards renal function is a little questionable, particularly in the light of subsequent work and in the context of more widely accepted use of high doses. In some of the studies no data on renal function were actually presented [8]; some used urine output only, with serial blood urea measurements, as an indication of renal function; in others, only a small number of patients was included [7] so a relatively low incidence of impaired renal function might easily have remained undetected. In none of the studies were controls established. This lack of proper control, in fact, dogs the debate about the incidence, and even the very existence, of contrast agent-associated nephrotoxicity up to the present day. In more recent years, the safety of contrast agents administered against a background of pre-existing renal failure has become less certain, following a considerable number of reports of deteriorating renal function associated with such administration. In some prospective studies, contrast media were reportedly involved in up to 10% of all cases of renal failure occurring in hospitalized patients, thus apparently exceeding aminoglycoside antibiotics in nephrotoxic potential. The incidence of reported renal problems has been increasing since 1970, perhaps as a result of the use of increasing doses, because of a heightened awareness of the problem and, paradoxically, with the tendency to undertake contrast studies with newer contrast agents on sicker patients than in the past.

4

However, problems arise in evaluating all these reports as much as in evaluating the earlier ones. They are retrospective and uncontrolled and, to complicate matters more, a wide range of definitions of contrast agent-induced nephropathy is used. Table 1 shows just some of the different degrees of serum creatinine accepted as the criteria for a diagnosis of contrast-associated nephropathy by various authors [10–15].

Table 1: Various criteria for nephrotoxic events

Reference	Serum creatinine concentration
Eisenberg et al [10]	1980: >1.0 mg/dl (90 µmol/L)
Carvallo et al [11]	1977: >2.0 mg/dl (180 µmol/L)
D'Elia et al [12]	1982: >1.0 mg/dl (90 µmol/L)
Byrd and Sherman [13]	1979: >2.0 mg/dl (180 µmol/L)
Older et al [14]	1980: >0.6 mg/dl (> 55 µmol/L) and at least 40% of precontrast level
Rao et al [15]	1980: >1.0 mg/dl (90 µmol/L)

Clinically, the glomerular filtration rate (GFR) is usually assessed indirectly by measuring serum creatinine concentrations or, more precisely, by measuring creatinine clearance. This definition may greatly underestimate toxicity not severe enough to affect such relatively insensitive markers of renal function. Serum creatinine concentration, which is used most often as an indicator of renal dysfunction, may not be elevated above the normal range until the GFR falls significantly below 50%. In the majority of reported cases, the serum creatinine level returns to baseline values within 7–10 days but renal failure, requiring short- or even long-term dialysis, is an outcome which occurs in some groups. An unchanged serum creatinine concentration cannot, of its nature, exclude any more subtle and long-term effects such as loss of functional capacity in the kidney and no one has yet studied whether these apparently transient adverse effects of contrast agents on renal function in any way affect the rate of progression towards endstage renal failure. It is certain that the true incidence of non-oliguric renal impairment associated with contrast agent administration is not known because it is by no means routine to monitor renal function systematically following contrast agent administration.

It is important to understand in any discussion of these matters that contrast agent-associated nephro-

toxicity is a very difficult concept to pin down. Patients in or visiting hospital and undergoing contrast-enhanced diagnostic and interventional procedures may be subject to a number of nephrotoxic insults: they may be dehydrated; they may be receiving other definitely nephrotoxic drugs; they may have already-impaired renal function and, unknown to their doctors, may be on a declining curve of renal function before contrast is administered. Just as hospitals are sometimes said, in general, to be dangerous for patients' health, so perhaps we may consider hospitals to be fundamentally nephrotoxic! Indeed, perhaps there is an important message in the only study which utilized controls. Patients attending for CT were randomly assigned to a contrast enhancement or non-contrast enhancement group and their renal function before and after the procedure was studied. The incidence of apparently procedure-related disturbance of renal function was the same in both groups.

Nevertheless, it is on such essentially anecdotal evidence that all our dogma in this area is built. Not only is it accepted as an article of faith by both radiologists and nephrologists that contrast agents possess a nephrotoxic potential, but associated increased risk factors are dogmatically cited: dehydration, pre-existing renal impairment, old age, diabetes [16]. Below, we discuss the relationship between intravascular iodinated contrast agents and the kidney and the known interactions between them. Thereby, possible mechanisms by which contrast agents might be expected adversely to affect renal function are indicated. We then briefly discuss the animal model data and the role of the modern low osmolality contrast agents. Finally, empirical approaches to the avoidance of contrast agent-associated renal problems are proposed.

CONTRAST HANDLING BY THE KIDNEY

Following intravascular injection, the iodinated contrast agents are rapidly distributed throughout the whole extracellular fluid space of the body, intra- and extravascular. The materials are passively filtered by the glomeruli and are excellent markers of glomerular filtration rate. There is no active secretion or reabsorption of contrast agent by the tubular cells. Less than 1% of administered agent is non-renally excreted if renal function is normal. Filtered contrast agent proceeding along the nephron is concentrated. In the proximal tubule some 85% of water in the filtrate is automatically reabsorbed. In the distal and collecting tubules, water reabsorption is under the control of antidiuretic hormone (ADH/vasopressin). In the presence of normal renal function and dehydration, very high concentrations indeed of contrast agent may be achieved in the distal tubules. These concentrating mechanisms are opposed by the osmotic effects produced by the contrast agents themselves, and the higher osmolality agents in high doses may overwhelm the concentrating mechanisms. This occurs more readily, and at lower doses, when, in patients with impaired renal function, the same contrast load is filtered by a diminished number of functioning glomeruli. The lower osmolality agents, both ionic and non-ionic, are only able to override the concentrating mechanisms in any given patient at a higher administered dose and, at any dose in a given patient, will be found in higher concentrations in the urine than the higher osmolality agents. In the past, and to some extent still, active dehydration was inflicted upon patients in order to obtain high urinary concentrations and denser pyelograms in intravenous urography. Most centres, however, have abandoned this practice because of the alleged association between dehydration and the adverse effects of contrast agents on the kidney [16–18].

PATHOPHYSIOLOGY OF ADVERSE CONTRAST EFFECTS IN THE KIDNEY

Since 99% of an intravascularly administered iodinated contrast medium is excreted by the renal route, the kidney may be legitimately seen as a target organ for these agents, more so since normal kidneys actively increase their concentration after filtration. Indeed, while the kidneys represent less than 0.5% of total body mass, they receive up to 25% of the resting cardiac output and have a high oxygen consumption, rendering them sensitive to any events or agents impairing oxygen uptake in any way. Furthermore, as indicated, it is one of the basic purposes of the kidney to excrete foreign and toxic substances while preserving water – in other words, to increase the concentrations of these materials as they pass through it. Patients with a reduced nephron population excrete the contrast material, or the foreign or toxic substances, through the smaller number of nephrons, thereby increasing the 'dose' per nephron. Volume depletion of patients or any obstruction to the flow of urine may lead to exposure to a greater concentration for a longer period.

Furthermore, since contrast media are excreted primarily by glomerular filtration, renal insufficiency results in a longer plasma half-life of the contrast medium and a greater exposure, therefore, for the kidney.

Immediately after administration of a contrast agent into the circulation the size of the kidneys changes. The decrease and subsequent increase are similar whether a high osmolar or low osmolar agent is used [19]. Within seconds, a powerful osmotic diuresis begins when high osmolar agents are used and a less marked diuresis following non-ionic agents [20]. Since the distension of the kidney is similar following both types of media the phenomenon seems unrelated to osmolality, and no entirely convincing explanation is currently available.

Renal perfusion

Immediately after the contrast agents reach the kidney there is an initial increase followed by a more prolonged decrease in renal blood flow. These changes are greater in dehydrated animals than in normally hydrated ones and greater following high osmolality agents than low osmolality agents [21]. The decrease in blood flow may be long-lasting, especially in the dehydrated animal, and may severely affect renal function. Indeed, during renal angiography in dogs, it has been shown that high osmolality agents may affect renal blood flow so severely as to produce a patchy contrast medium retention effect [22]. What appears to be the same phenomenon has been observed by Love et al [23] in some patients having CT, 24 hours or so after an arteriogram (Figure 1).

The effect appears in Love's work [24] to be correlated with degree of impairment of renal function. Indeed, these observations are, in this writer's opinion, far more convincing than the wealth of other, largely anecdotal case reports, based on measurements of creatinine, that contrast agents really do have a nephrotoxic potential of clinical importance.

6

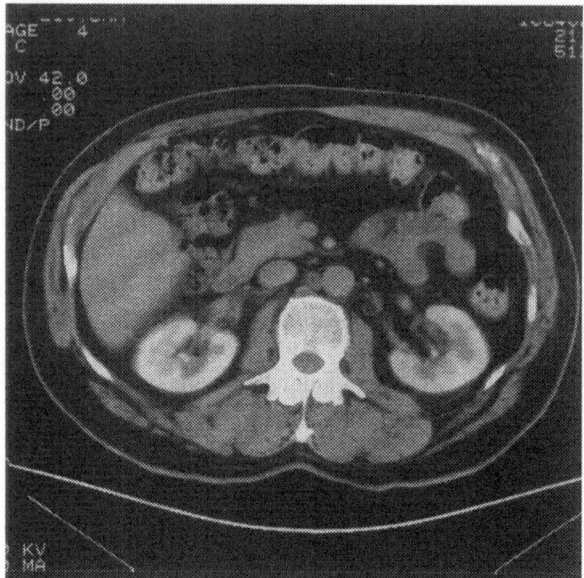

Figure 1 CT scan the day after angiography. Notice the renal cortical enhancement though no contrast has been given for more than 12 hours. (Images kindly loaned by Dr. Leon Love of Loyola Medical Centre.)

Glomerular injury

Trewhella et al [17,18] have reported that contrast agents administered as rapid bolus injections result in a rapid rise in circulating vasopressin levels, which has an adverse effect on renal perfusion (Figure 2).

In dog experiments, not only are vasopressin concentrations appropriately elevated in dehydration but the administration of materials such as contrast agents, increasing the osmolality of body fluids, produces an exaggerated vasopressin rise response. The authors suggested that vasopressin is consequently an important mediator of contrast agent-associated nephrotoxic effects. The hypothesis would certainly explain the alleged association with dehydration as a risk factor.

Other factors which may impair renal blood flow include the well-recognized red cell rigidification engendered by contrast agents [3]. Table 2 summarizes various levels at which contrast agents have, or may be expected to have, adverse effects on the kidney.

When contrast agents reach the glomeruli they undoubtedly are capable of producing injury since a

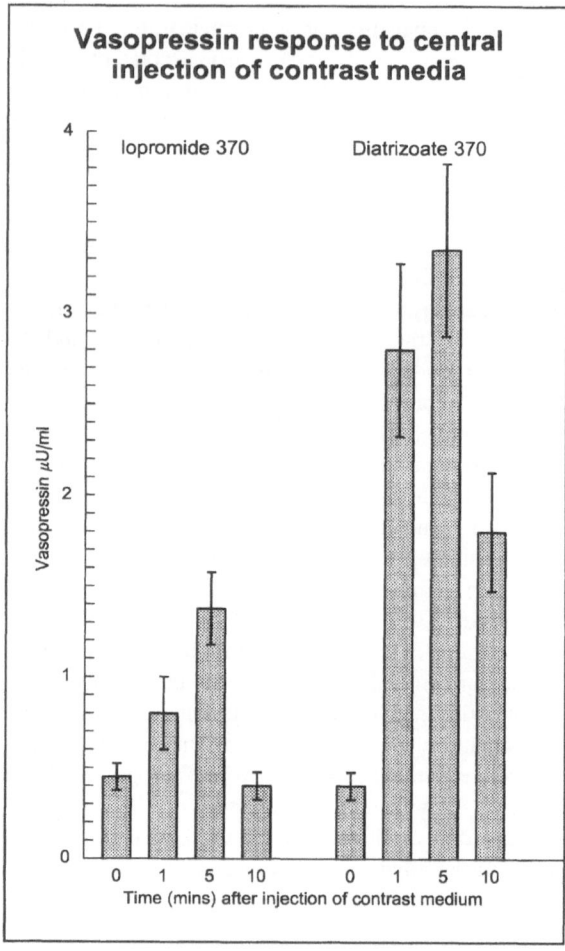

Figure 2 Vasopressin responses in normally hydrated patients to bolus administration of 'high' [Diatrizoate] and 'low' [Iopromide] osmolality contrast agents. The rises associated with the 'high' osmolality agent are from normal range to dehydration ranges. Much higher base levels and larger rises could be anticipated in the dehydrated patient.

glomerular protein leak is seen [22,25–27] and, in this sense, are undoubtedly nephrotoxic. The effect is greater with the higher osmolality agents than with the lower osmolality agents but osmolality is clearly not the only factor since metrizamide produces much the same degree of albuminuria as do the high osmolality agents.

Table 2: Some possible mechanisms of contrast nephrotoxicity

(a)	Impairment of Perfusion	(i)	Vasopressin-mediated
		(ii)	Red cell rigidification effects
		(iii)	Intrarenal pressure changes
		(iv)	Systemic hypotension
(b)	Glomerular Injury	(i)	Osmolality-mediated
		(ii)	Chemotoxicity-mediated
		(iii)	Hypoxic
(c)	Tubular Injury	(i)	Osmolality-mediated
		(ii)	Chemotoxicity-mediated
		(iii)	Hypoxic
(d)	Obstructive Nephropathy	(i)	'Osmotic nephrosis'
		(ii)	Precipitation of contrast agents with:
			(a) Normal urinary proteins
			(b) Abnormal urinary proteins
			(c) Tubular cellular debris

TUBULAR CELL INJURY

When descending along the nephron and, in the normal kidney, being concentrated dramatically, contrast agents are clearly capable of injuring the tubular cells since a variety of tubular cell-specific and tubular brush border enzymes may be seen in the urine [25,28,29]. This again clearly illustrates a potential for renal injury by contrast agents and, equally clearly, has to be labelled nephrotoxicity. However, the relationship of such routine events to the occasional episodes in a few patients of significant changes in renal function is unclear. Nevertheless, not surprisingly, a large literature has been spawned on this subject. One of the findings has been that, rather like the glomerular damage, high osmolality is involved but is not the exclusive mechanism. Low osmolality agents still produce the phenomenon, though to a lesser degree and not in simple proportion to their osmolality. It may be that the magnitude of the effect may be exacerbated by impairment of renal blood flow and tubular cell hypoxia.

OBSTRUCTIVE NEPHROPATHY

Since, occasionally, a persistent dense nephrogram is said to be associated with impaired renal function following contrast administration (Figure 1), it has been suggested that the whole phenomenon might be based on an obstructive nephropathy. Contrast agents might precipitate with Tamm–Horsfall protein, with abnormal proteins in, for example, myeloma, or with proteinaceous material resulting from the above-discussed injury to tubular cells. The evidence for this, however, is slim. Saxton and colleagues were unable to obtain precipitation of contrast agents with abnormal myeloma proteins and Dawson and colleagues were unable to demonstrate precipitation with Tamm–Horsfall protein [16]. Microscopic changes are seen after the administration of both ionic and non-ionic contrast agent in tubular cells and have been described as osmotic nephrosis [29]. This is a misnomer since it appears to be largely independent of osmosis. The vacuoles seen in the tubular cells seem to be pinocytosed contrast agent. It was once thought that the swelling of tubular cells associated with widespread osmotic nephrosis might itself result in an obstructive nephropathy and explain a dense persistent nephrogram. However, the effect, once again, appears to be routine rather than idiosyncratic and the hypothesis has not stood up to the test of time and further study.

ANIMAL MODELS

A great variety of animal models has been studied and some of these and their uncertainties have been discussed by Thomsen et al [30]. One particular study, by Vaamonde et al [31], was an exhaustive one in the rat and, most interestingly, failed to elicit any convincing evidence of nephrotoxic effects other than proteinuria in animals with 'a remnant kidney' (pre-existing renal dysfunction model), but this was not associated with any significant degree of renal dysfunction. This brings us back to the point about tubular injury undoubtedly being an aspect of nephrotoxicity but not being in any way obviously related to clinically important events. This striking failure in carefully conducted experiments in a reasonable model moved Katzberg, in an editorial [32], to suggest that "there is a real possibility that the 'null hypothesis' may represent the truth". That is to say, there may be no such thing as genuine contrast agent-associated nephrotoxicity! This may be too dogmatic a conclusion, but the failure to develop a convincing animal model for contrast agent-associated nephrotoxicity is salutary. We should remember the very largely anecdotal nature of the evidence on which our beliefs are based.

NEW CONTRAST AGENTS

The introduction, in the 1980s, of low osmolality contrast agents (ionic and non-ionic) has served to muddy the waters somewhat [2,3]. The differences in toxicity between the conventional ionic, low osmolality ionic and low osmolality non-ionic agents have been reviewed elsewhere [3]. Since the non-ionic agents exhibit a significantly lower toxicity than other agents, in general terms, it is reasonable, from first principles, to postulate that they might be associated with a lower incidence of associated nephrotoxicity. Three studies [33–35] claim to have demonstrated no significant differences in clinical practice between non-ionic and conventional agents in this regard, contrary to theoretical speculations. However, these studies are open to a number of very serious criticisms. Schwab et al [34] and Parfrey et al [33] examined only transient changes in serum creatinine and neither encountered major acute renal failure. Gomes et al [35] examined only 'high-risk' patients and found no statistically significant differences in the incidence of renal dysfunction between groups of patients receiving ionic and non-ionic contrast material. However, a crucial fact may be that 5 out of 290 patients required dialysis and all were in the ionic group. It seems reasonable to suggest that insufficient numbers of 'high-risk' patients were studied to allow comment on the risk of clinically important phenomena. In any case, as was perhaps ethically necessary, all patients in the three studies were carefully hydrated prior to contrast administration, thereby eliminating what is widely held to be the single most important risk factor for this phenomenon. These studies were, for this and other reasons, inadequate to the task of eliciting differences between ionic and non-ionic contrast agents as regards threat to renal function. Katholi et al [36], on the other hand, compared non-ionic low osmolality and ionic high osmolality contrast agents in a prospective, double-blind randomized study of 70 patients, with normal or mildly depressed renal function, undergoing coronary angiography. They determined creatinine clearance at 24 and 48 hours after the procedure. There were no significant differences between the low and high osmolality groups with regard to age, baseline creatinine clearance or dose of contrast medium given, but in the patients receiving low osmolality agents creatinine clearance decreased by 19% compared with 40% in the patients receiving high osmolality medium. It was concluded that this was some evidence, albeit in a small series, of the greater safety in this regard of non-ionic agents.

If the jury is still out on the true incidence and significance (not to mention the very existence) of contrast agent-associated nephrotoxicity, it is not surprising that it is still out on differences between different kinds of agent.

CONCLUSIONS

The assumption that iodinated intravascular contrast agents have a nephrotoxic potential has been with us for half a century and the subject of serious study for a quarter of a century. There is an enormous amount of literature on the subject but it only allows us certain limited conclusions: contrast agents can indeed cause renal injury to glomeruli and tubules which results, in animals and man, in heavy proteinuria. The relationship between this and clinically important renal dysfunction has not been demonstrated; non-ionic, second-generation contrast agents are less toxic in this regard but, once again, the clinical relevance of this is by no means clear; many patients exhibit a rise in serum creatinine in association with contrast administration but this is transient and no relationship between this common event and the more dramatic manifestations of renal impairment has been demonstrated. All so-called studies have been bedevilled by lack of controls to eliminate the numerous other nephrotoxic influences on patients receiving contrast and by the variety of definitions of contrast agent-associated nephrotoxicity adopted; specific contributions to alleged risk, such as by pre-existing renal impairment, old age, diabetes and dehydration, have not been definitively established. All our so-called knowledge is based on anecdotal and uncontrolled evidence.

As for practical implications, we may reasonably say:
(1) No contrast agent is best.
(2) Low dose is better than high dose.

(3) Dehydration should be corrected, wherever possible, before any procedure and active dehydration for urography abandoned.

(4) The evidence that non-ionic agents are safer in this regard has not been conclusively demonstrated but, on the basis of their established lower toxicity in all other respects, their use is strongly recommended, at least for all patients with pre-existing renal or cardiac dysfunction.

As scientists, we should be sceptical and remember Katzberg's injunction about the 'null hypothesis'. As clinicians, we owe a duty of care to the patient and must exercise caution.

REFERENCES

1. Grainger RG. Intravascular contrast media – the past, the present and the future. Br J Radiol. 1982; 55: 1–18.
2. Dawson P, Grainger RG, Pitfield J. The new low osmolality contrast agents. A simple guide. Clin Radiol. 1983; 34: 221–226.
3. Dawson P. Chemotoxicity of contrast media and clinical adverse effects: a review. Invest Radiol. 1985; 20: 52–59.
4. Macewan DW, Dunbar JS, Nogrady MB. Intravenous pyelography in children with renal insufficiency. Radiology. 1962; 78: 893–903.
5. Schwartz RH, Hurwitt A, Ettinger A. Intravenous urography in the patient with renal insufficiency. N Engl J Med. 1963; 269: 277–283.
6. Gup AD, Fischman JL, Aldridge G, Schlegal JK. The effect of drip in infusion pyelography on renal function. Am J Roentgenol. 1966; 98: 102–106.
7. Bengtsson U, Cederbom G, Falkheden T, Jagenburg R. Can J Urol Nephrol. 1968; 2: 173–176.
8. Fry IK, Cattell WR. Excretion urography in advanced renal failure. Br J Radiol. 1971; 44: 198–202.
9. Davidson AJ, Becker J, Rothfield N, Unger G, Ploch DR. An evaluation of the effect of high dose urography on previously impaired renal and hepatic function in man. Radiology. 1970; 97: 249–254.
10. Eisenberg RL, Bank WO, Hedgcock MW. Renal failure after major angiography. Am J Med. 1980; 68: 43–46.
11. Carvallo A, Rakolwski TA, Argy WP, Schreiner GE. Acute renal failure following drip infusion pyelography. Am J Med. 1978; 65: 38–45.
12. D'Elia JA, Gleason RE, Aldy M. Nephrotoxicity from angiographic contrast material. A prospective study. Am J Med. 1982; 72: 719–725.
13. Byrd L, Sherman SY. Radiocontrast-induced acute renal failure: A clinical and pathophysiologic review. Medicine. 1979; 58: 270–279.
14. Older RA, Korobkin M, Cleeve DM, Schaaf R, Thompson W. Contrast-induced acute renal failure: Persistent nephrogram as clue to early detection. Am J Roentgenol. 1980; 134: 339–342.
15. Rao SR, Miexa A, Leiter E. Renal failure in diabetes after intravenous urography. Urology. 1980; 15: 577–580.
16. Dawson P. Contrast agent nephrotoxicity: an appraisal. Br J Radiol. 1985; 58: 121–124.
17. Trewhella M, Forsling M, Richards D, Dawson P. Dehydration, antidiuretic hormone and the intravenous urogram. Br J Radiol. 1987; 60: 445–447.
18. Trewhella M, Dawson P, Forsling M, McCarthy P, O'Donnell C. Vasopressin release in response to intravenously injected contrast media. Br J Radiol. 1990; 63: 97–100.
19. Whitehouse RW. High- and low-osmolar contrast agents in urography. A comparison of the appearances with respect to pyelotubular opacification and renal length. Clin Radiol. 1986; 37: 395–398.
20. Sjöberg S, Almen T, Golman K. Excretion of urographic contrast media. Iohexol and other media during free urine flow in the rabbit. Acta Radiol. 1980; 362 (Suppl): 93–98.
21. Katzberg RW, Morris TM, Lasser EC, et al. Acute systemic and renal hemodynamic effects of meglumine/sodium diatrizoate 76% and iopamidol in euvolemic and dehydrated dogs. Invest Radiol. 1983; 21: 793–797.
22. Törnquist C. Nephrotoxicity of ionic and non-ionic contrast media in experimental and clinical nephroangiography. With special reference to ionic diatrizoate and metrizoate and non-ionic iohexol. 1985. Thesis, University of Lund.
23. Love L, Lind JR, Olson MC. Persistent CT nephrogram: significant in the diagnosis of contrast nephropathy. Radiology. 1989; 172: 125–129.
24. Love L. 1993. Private communication.
25. Holtas S. Proteinuria following nephroangiography. 1978. Thesis, Malmo General Hospital, Malmo, Sweden.
26. Thomsen HS, Dorph S, Mygind T, et al. Intravenous injection of ioxilan, iohexol and diatrizoate. Effects on urine profiles in the rat. Acta Radiol. 1988; 29: 131–136.
27. Thomsen HS, Hemmingsen L, Dorph S, Skaarup P. Effects on urine profiles of idatrizoate in hydrated and dehydrated rats. A double cross-over study. Acta Radiol. 1988; 29: 731–735.
28. Goldstein EJ, Feinfield DA, Fleischner CM, Elkin M. Enzymatic evidence of renal tubular damage following renal angiography. Radiology. 1976; 121: 617–619.
29. Moreau JF, Droz D, Noel LH, Heibowitch J, Ungers P, Michel JR. Tubular nephrotoxicity of water soluble iodinated contrast media. Invest Radiol. 1980; 15: S54–S60.
30. Thomsen HS, Golman K, Hemmingsen L, Skaarup P, Svendsen O. Contrast medium induced nephropathy animal experiments. Front Eur Radiol.1993; 9: 87–107.
31. Vaamonde CA, Bier RT, Papendick R, Alpert H, Gouvea W, Owens B, et al. Acute and chronic renal effects of radiocontrast in diabetic rats. Role of anaesthesia and risk factors. Invest Radiol. 1989; 24: 210–218.
32 Katzberg RW. What do we really know about contrast medium-induced acute renal failure? Invest Radiol. 1989; 24: 219–220.
33. Parfrey PS, Griffiths SM, Barrett BJ, Paul MD, Genge M, et al. Contrast material induced renal failure in patients with diabetes mellitus, renal insufficiency, or both. N Engl J Med. 1989; 320: 143–149.

34. Schwab SJ, Hlatky MA, Pieper KS, Davidson CJ, Morris KG, Stelton TN, et al. Contrast nephrotoxicity: a randomized controlled trial of a non-ionic and an ionic radiographic contrast agent. N Engl J Med. 1989; 320:149–153.
35. Gomes AS, Lois JF, Baker JD, McGlade CT, Bunnell DH, Hartzman S. Acute renal dysfunction in high-risk patients after angiography: Comparison of ionic and non-ionic contrast media. Radiology. 1989; 170: 65–68.
36. Katholi RE, Taylor GJ, Woods WT, et al. Nephrotoxicity of non-ionic low-osmolality versus ionic high osmolality contrast media: a prospective double blind randomized comparison in human beings. Radiology. 1993; 186: 183–187.

This paper first appeared in *Advances in X-Ray Contrast*. 1993;1:2–9.

UPDATE

As indicated in the article, careful analysis of the available data shows it leaves much to be desired. There is virtually no work utilising proper controls, the relationship of some of the parameters measured to clinically important changes in renal function has not been clearly established, and no reliable animal model exists for further exploration. This may be thought surprising, and not a little disappointing, given the fact that iodinated X-ray contrast agents have been in use for more than 60 years. That the question of whether non-ionics are safer in terms of renal events than ionics has not been answered seems unsurprising in view of this background.

P. Dawson and W. Clauss, (eds.), Advances in X-Ray Contrast: Collected Papers. 11–19
© *1998 Kluwer Academic Publishers.*

Nephrotoxicity related to X-ray contrast media

Knut Joachim Berg[1] and Jarl Å. Jakobsen[2]
[1]*Section for Nephrology, Medical Department B and* [2]*Department of Radiology, The National Hospital,*
Rikshospitalet, Oslo, Norway

ABSTRACT

This survey discusses pharmacokinetic aspects and the renal glomerular and tubular side-effects of modern low osmolar X-ray contrast media (LOCM). The clearance of X-ray contrast media (CM) approximates to the glomerular filtration rate (GFR) and the CM are concentrated during their passage through the renal tubular system up to 100 times the plasma concentration during peak diuresis in patients with normal GFR. The urinary CM concentration is relatively lower in patients with renal failure.

Methods for evaluation of GFR and renal tubular function are discussed. Serum creatinine is a rather insensitive parameter for estimation of GFR, especially in patients with a small reduction of GFR, and plasma or renal clearance methods are recommended. However, serum creatinine reflects major drug-induced fluctuations in GFR, and is a suitable parameter for determination of GFR in most instances. Serum creatinine often peaks more than 72 hours after administration of CM in patients with reduced GFR, and should be monitored for at least 5 days in such patients.

LOCM seem to be less toxic when administered intravenously than when given intra-arterially. However, from randomized trials one can conclude that LOCM reduce the incidence of acute renal failure, as defined as an increase in serum creatinine, especially in high-risk patients.

A dose-dependent increase in the urinary excretion of renal tubular lysosomal and brush border enzymes is regularly observed after administration of CM. Dimeric LOCM seem to affect the excretion of lysosomal enzymes less than monomeric LOCM. Enzymuria is also observed after intravenous administration of equimolar doses of mannitol, but urinary enzyme excretion is significantly greater and more long-lived after CM. Thus, the enzymuria cannot be explained only by osmotic effects of the CM.

The clinical importance of this enzymuria is discussed.

PHARMACOKINETICS

X-ray contrast media (CM) are small molecules with low protein binding that are freely filtered through the glomerular basement membrane. As renal tubular transport of CM is negligible, they behave more or less like inulin in the kidney, and both plasma and renal clearances of CM may be used for determination of glomerular filtration rate (GFR). The serum concentration of CM varies with the dose administered, time after injection and GFR. In subjects with normal GFR, the serum concentration of the non-ionic CM, iohexol [1] and iopentol [2], is 5–10 mg/ml (6–12 mmol/L) 1–2 hours after administration of 1.2 g I/kg b.w. Thereafter, the concentration declines exponentially, with a mean elimination half life ($t_{1/2\beta}$) of approximately 120 min. Seventy-five percent of the dose is excreted in urine within 4 hours, so the urinary CM concentration declines rapidly. As 99% of ultrafiltered water is reabsorbed in the kidney, CM is concentrated 100 times in the urine of patients with normal GFR, with a peak of 200–500 mg/ml (240–600 mmol/L) during the first 4-hour period after administration of high doses. Urine is concentrated only 4-fold during its passage through the proximal tubule, so the peak concentration in the late proximal tubule is approximately 4 times the plasma concentration. Consequently, the osmotic load presented to the kidney tubules during the first hours after injection is very high, even after administration of non-ionic monomeric CM, and comparable with the concentrations used in in vitro experiments with rabbit proximal tubule segments [3], cell cultures of renal epithelial cell [4], or other cell lines [5].

12

In patients with impaired renal function the urinary excretion of CM is protracted and the elimination half-life inversely correlated to GFR. The half-life of iopentol was 28.4 (6.8–69.2) hours in a group of patients with a mean GFR of 9.3 ml/min, about 14 times the half-life in healthy volunteers [6]. The initial concentration of CM in urine was much lower than in patients with normal GFR (20 versus 200–500 mg/ml during the first 0–6 hours), but CM could be detected in urine for 5–7 days after injection. This means that the acute osmotic load presented to the kidney is much lower, as is probably also the load to the functioning tubules, but the effect of the CM lasts longer. It is, therefore, important to follow renal function parameters for as long as 5–7 days after CM administration in patients with reduced GFR.

EFFECTS OF CM ON GLOMERULAR FILTRATION RATE

CM may affect both renal glomerular and tubular function, as indicated by changes of GFR parameters, renal excretion of proteins such as albumin and microproteins, and changes in urinary enzyme excretion rate. The incidence of CM-associated nephropathy is, of course, dependent on the diagnostic criteria used. It is generally accepted that the diagnosis of CM nephropathy is primarily related to the effect of CM on GFR. The methods used for the assessment of GFR are, therefore, important.

GFR parameters

Ideally, GFR should be controlled by a clearance method, such as *plasma clearance* of an isotope (^{51}CrEDTA, ^{99m}TcDTPA) or of CM. Pharmaco-kinetic determination of plasma clearance with isotopes usually involves plasma sampling for 3–4 hours, and overestimates GFR in patients with reduced renal function. In such patients, a 24-hour plasma sample should be included. Plasma clearance of isotopes is far more reproducible than *creatinine clearance* and should be performed, when possible, for detailed control of GFR in small population studies. In patients with impaired renal function, creatinine clearance overestimates GFR because creatinine is secreted in the renal tubule. The coefficient of variation

of creatinine clearance determination is also critically dependent on the urine sampling and is greater than that for *serum creatinine*. Serum creatinine is, however, often an unreliable parameter for estimation of the GFR in patients with a small reduction in GFR (e.g. elderly people) as GFR may be reduced by as much as 50% before serum creatinine is significantly increased. In order to validate absolute GFR levels in such patients, one can use a formula for *calculated creatinine clearance*, which includes serum creatinine, age and body weight.

For practical purposes, one often has to rely on *serum creatinine* as a standard indicator for estimation of GFR. Serum creatinine reflects drug-induced fluctuations in GFR, and is a suitable parameter for determinations of short-term effects of CM in most instances. Serum creatinine is often increased within 24 hours and usually peaks 48–72 hours after administration of CM, or even later in patients with reduced GFR. It is, therefore, important to control the GFR parameter after 5–7 days in patients with impaired renal function and in other risk patients.

The diagnosis of CM nephropathy

This is most often based on changes in serum creatinine levels. A minor effect on GFR is often defined as an increase in serum creatinine of 44 µmol/L (0.5 mg/dl), and a major effect as an increase of 88 µmol/L (1 mg/dl) or more. Of the patients who display major effects on GFR, only a small percentage (10–15%) develop acute oliguric renal failure.

The incidence of CM-induced nephropathy

This varies widely according to different reports and is dependent on individual patient risk factors. Intravenous contrast administration represents a very low risk (less than 1%) for acute renal failure in healthy people. The two most important risk factors are reduced renal function, as indicated by an increased serum creatinine level before the investigation, and diabetes mellitus [7–10]. The risk of developing CM nephropathy after femoral angiography was shown in one study to be as follows: 2% in low-risk patients (serum creatinine <133µmol/L, no diabetes), 16% in diabetics with serum creatinine <133 µmol/L, 22% in

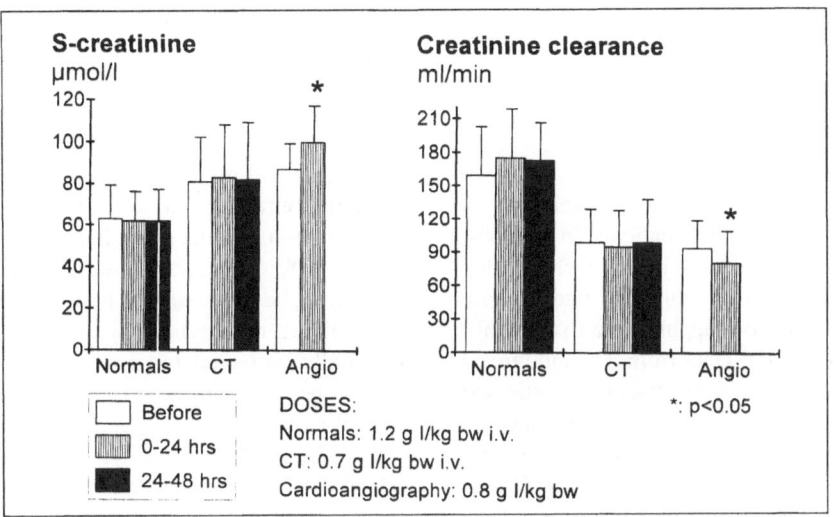

Figure 1 Serum creatinine and creatinine clearance after injection of the non-ionic monomeric CM iopentol (Nycomed AS, Oslo, Norway). After i.v. injection in healthy volunteers (normals, $n=8$), and in patients for CT ($n=32$), no significant changes occurred. After cardioangiography (angio, $n=30$), there was a reduction in GFR. The doses were highest in the healthy volunteers (1.2 g I/kg body weight (b.w.)), while they were nearly similar in the two patient groups. Vertical lines represent 1 standard deviation. Change from baseline with $p<0.05$ indicated by * [12].

patients with serum creatinine >133 μmol/L and no diabetes, 38% in diabetics with serum creatinine >133 μmol/L [9]. In another study, in 59 high-risk diabetic patients with creatinine clearance levels <30 ml/min [11], nine patients required dialysis within 14 days of contrast administration because of reversible CM-induced acute renal failure. High dose, dehydration, low arterial pressure, and heart failure, as indicated by a low ejection fraction, were also risk factors in these patients. Other risk factors, which may be added to this list, are old age, repeated CM administration, bowel purgation by enema and co-administration of other nephrotoxic drugs.

Intravenous versus intra-arterial administration

In controlled, randomized studies of iopentol in healthy volunteers and in patients with normal renal function, the CM seemed to be less toxic when administered i.v. than when given i.a. for cardioangiography. Serum creatinine was increased and creatinine clearance reduced after i.a. contrast administration, indicating a significant reduction of GFR, whereas the parameters were unchanged after i.v. contrast administration [12] (Figure 1). Our studies thus seem to indicate that low osmolar CM (LOCM) are more nephrotoxic when given i.a. than when given i.v.

High osmolar CM (HOCM) also seem to be more nephrotoxic after i.a. than after i.v. administration. In a retrospective study, acute renal failure, defined as an increase in serum creatinine > 2 mg/dl (>176 μmol/L), occurred in 0.53% of the patients after angiography compared with 0.15% after i.v. urography or i.v. CT [13]. In another, prospective, double-blind study of iohexol, iopamidol and the dimeric ionic CM, ioxaglate, serum creatinine tended to increase more after i.a. than after i.v. injection [14]. In a large meta-analysis of the relative toxicities of HOCM and LOCM [15], LOCM were found to be less likely to prevent CM nephropathy after i.v. than after i.a. injection. Significant effects of LOCM could only be demonstrated after i.a. contrast administration.

Taking into account the fact that the acute intrarenal CM concentrations are higher after suprarenal i.a. than after i.v. contrast administration, these observations are not surprising.

Is there any difference in nephrotoxicity between LOCM and HOCM?

As expected, it is difficult to detect differences in nephrotoxicity of CM in low-risk patients. In high-risk patients, most, but not all, studies have concluded that monomeric and dimeric non-ionic CM and dimeric ionic CM (LOCM) are less nephrotoxic than monomeric ionic CM (HOCM). Data from 31 randomized trials including more than 5000 subjects investigated with LOCM and HOCM have recently been collected and the results have been pooled [15]. Considering all these trials together, there was a statistically significant (p=0.02) smaller decline in renal function after LOCM than after HOCM. Among 25 trials with available data, the pooled odds of a rise in serum creatinine after LOCM was 0.61 times that after HOCM. Large changes in serum creatinine occurred only in patients with existing renal failure, and were less common with LOCM. The authors conclude that LOCM are less nephrotoxic than HOCM in patients with existing renal failure. The data also indicate that the patients benefit more from LOCM after i.a. than after i.v. administration.

It is also reasonable to postulate that LOCM are less nephrotoxic than HOCM not only in patients with reduced GFR and/or diabetes mellitus, but also in other high-risk patients.

EFFECTS OF CM ON RENAL TUBULAR FUNCTION

Determination of urinary enzymes

Urinary excretion of renal tubular enzymes and tubular antigens appears to be a sensitive tool for the identification of nephrotoxic drugs. Ideally, the enzymes investigated should fulfil the following criteria [16]:
(1) high concentration in kidney parenchyma,
(2) large M.W. (no glomerular filtration),
(3) stable for several days (in refrigerator or freezer),
(4) inhibitors absent in urine,
(5) their determination should be accurate and valid.

Unfortunately, many enzymes are unstable when stored and, in some cases, are also reduced in activity by enzyme inhibitors present in urine. Such inhibitors can be removed by gel filtration or dialysis of the urine, or have their effect reduced by addition of stabilizing agents such as glycerol to the urine sample.

When enzymes are collected for large-scale CM investigations, urine often has to be frozen for some weeks before being analysed. Normal values should, therefore, be presented for both fresh and frozen urine (e.g. after 2 months). Urinary excretion of enzymes is normally corrected for urinary creatinine concentration, whether the urine is sampled for 24 hours or not. Small children and old people excrete less creatinine than others, hence the enzyme ratio in these groups is normally increased. Because of diurnal variations in enzyme and creatinine excretion, one should always collect spot urine samples at the same time-points each day.

The renal tubular brush border enzymes, alkaline phosphatase (ALP) and alanine aminopeptidase (AAP) and the tubular lysosomal enzyme, N-acetyl-β-glucosaminidase (NAG), are sensitive parameters for proximal tubular function, and are the ones most widely used for validation of CM. NAG is less influenced by inhibitors than AAP and is more stable than ALP after freezing. Many other enzymes, for instance other lysosomal, cytoplasmic and brush border enzymes, as well as brush border-related proteins (antigens) [17], are also used.

Urinary β_2-microglobulin

This has been used as a parameter for proximal tubular function. The protein is freely filtered in the glomeruli, but 99.9% is normally reabsorbed in the proximal tubule, provided that the serum value of the protein does not exceed the renal threshold value for the tubular reabsorption. The urinary excretion of β_2-microglobulin is usually normal after administration of CM, but may be considerably increased in individual patients.

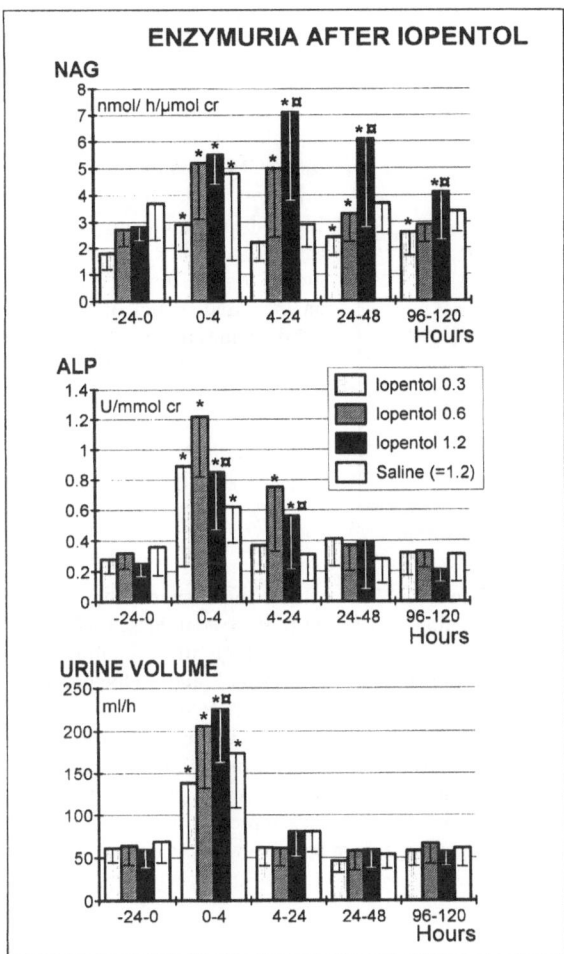

Figure 2 Changes in the renal tubular enzymes *N*-acetyl-β-glucosaminidase (NAG) and alkaline phosphatase (ALP) after i.v. injection of increasing doses of iopentol. The CM was given to healthy volunteers, *n*=8 in each group. The same volume of saline was given as of CM in the highest dose group. A dose-dependent increase in enzyme excretion, especially for NAG, is seen. A minor, short-lived increase is also seen after saline. Vertical lines represent 1 standard deviation. Change from baseline with *p*<0.05 indicated by * [18].

How is renal enzyme excretion affected by CM?

Urinary excretion of renal tubular enzymes (NAG, AAP, ALP) is usually increased after both i.v. and i.a. CM administration. Although non-ionic monomeric LOCM seem to affect urinary enzyme excretion somewhat less than the ionic HOCM [17], the excretion of NAG and ALP has been shown to be increased by LOCM, and in a dose-dependent manner [18] (Figure 2).

The excretion of these enzymes usually peaks 24 hours after administration of CM and then normalizes within 48 hours in patients with normal or nearly normal GFR (serum creatinine <120 μmol/L). The brush border enzymes seem to be more sensitive than NAG to CM, as they normally increase by 200–400%, compared with an increase in NAG excretion of 75–200%. The excretion of brush border enzymes is also somewhat more short-lived. This probably reflects a higher initial concentration of CM at the brush border site. Urinary NAG excretion and the excretion of kidney antigens was nearly doubled and significantly increased in 10 healthy volunteers after administration of 30 ml iohexol (300 mg I/ml) for determination of X-ray contrast clearance (BG Danielson, personal communication). The effect of i.v. and i.a. CM injection on urinary enzymes is of the same order, in contrast to the different effect of the two procedures on GFR [12] (Figure 3). Although an inverse correlation between changes in GFR and urinary enzyme excretion has been reported [17], such a correlation has not been found by others [12].

Is the CM-induced increase in urinary enzyme excretion merely an effect of the high osmotic load in the proximal tubule, or an indication of a direct tubulotoxic effect induced by the CM? In order to investigate this we administered equimolar doses of mannitol and CM (iopentol and the dimeric non-ionic, CM iodixanol) i.v. in four groups of healthy volunteers [19] (Figure 4).

Mannitol increased excretion of NAG and ALP during the 0 to 4-hour period after administration, i.e. during peak diuresis. But the CM increased ALP significantly more, their effects being still evident after 24–48 hours. Iodixanol had no effect on urinary NAG excretion, but iopentol increased NAG more than did mannitol. We conclude that the early effect of CM on urinary enzyme excretion can be related to increased diuresis, but that the late effect, apparent after 24–48 hours, cannot be explained by the osmotic load of the CM and is most probably caused by iodine or other specific CM-related factors.

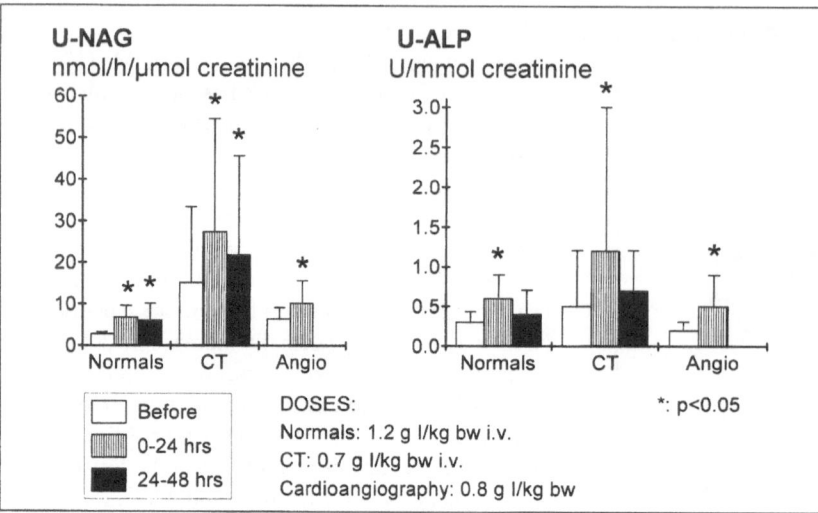

Figure 3 Changes in the renal tubular enzymes *N*-acetyl-β-glucosaminidase (NAG) and alkaline phosphatase (ALP) after i.v. injection of the non-ionic monomeric CM iopentol. Both after i.v. injection in healthy volunteers (normals, *n*=8), in patients for CT (*n*=30), and after cardioangiography (angio, *n*=24), there was an increased output of enzymes. Vertical lines represent 1 standard deviation. Change from baseline with *p*<0.05 indicated by * [12].

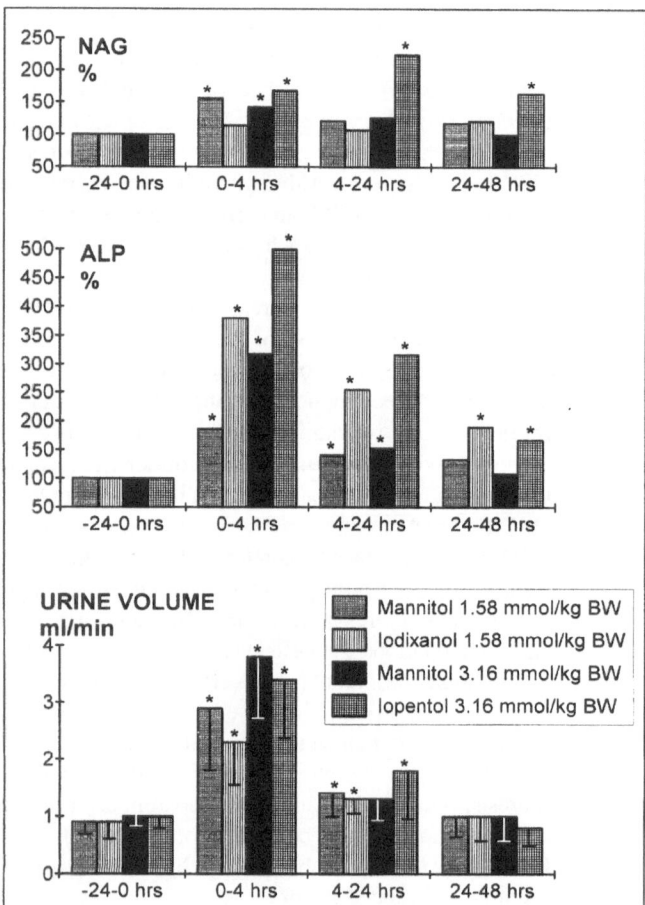

Figure 4 Urine volume and changes in the urinary excretion of *N*-acetyl-β-glucosaminidase (NAG) and alkaline phosphatase (ALP) after mannitol and equimolar doses of iopentol and iodixanol (Nycomed AS, Oslo, Norway). NAG and ALP were increased more after injection of CM and the enzymuria lasted longer than following mannitol administration (*n*=10, in each group). For urine volume, vertical lines represent 1 standard deviation. Change from baseline with *p*<0.05 indicated by * [19].

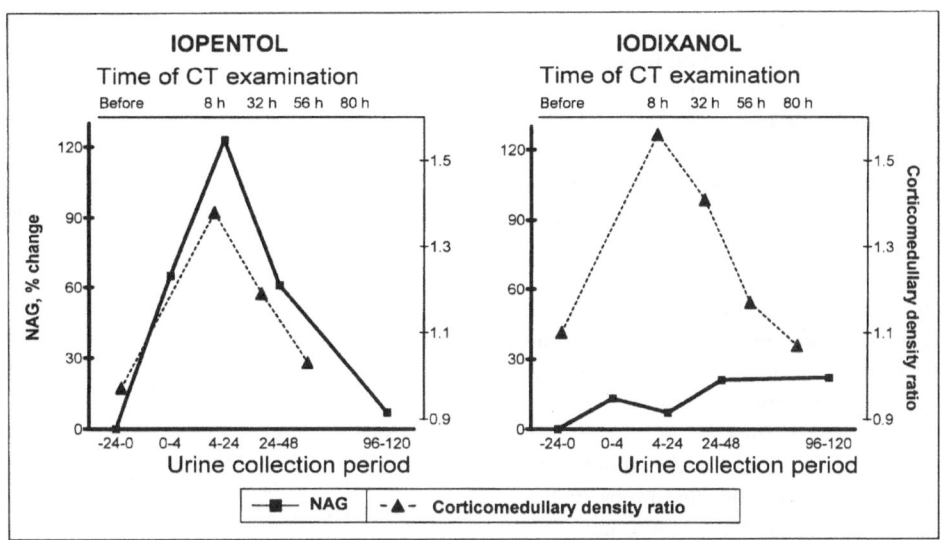

Figure 5 Changes in the urinary excretion of *N*-acetyl-β-glucosaminidase (NAG) (left axis), and in the corticomedullary density ratio (retention measured by CT) (right axis) after injection of 1.2 g I/kg b.w. of iodixanol and iopentol in healthy volunteers (*n*=10 in each group). Note that the lysosomal enzyme NAG is unrelated to the retention of CM.

After injection of CM in rats and mice, marked vacuolization of the proximal tubule, so-called 'osmotic nephrosis', has been demonstrated [20]. The vacuoles are seen after both ionic and non-ionic CM. Iodixanol induced more vacuolization than did iopentol. The content of the vacuoles is probably CM that have entered the cells by endocytosis [4,5]. The histological findings of 'osmotic nephrosis' correlated with the increase in urinary NAG excretion found in rats [21]. This vacuolization is also seen in man [22], and the vacuoles could possibly be related to CM retention evident on nephrograms, as demonstrated by CT. Since these vacuoles are localized in close proximity to the lysosomes, we examined the correlation between the cortical CM-related enhancement seen by CT after i.v. iopentol or iodixanol with urinary enzyme excretion [23] (Figure 5).

There was no correlation between the cortical enhancement shown on CT and the enzyme excretion. It is remarkable that iodixanol exerts a smaller effect on the lysosomal enzyme, NAG, than all other LOCM which we have investigated [23–25], in spite of the lysosome-related vacuolization observed.

Is CM-induced enzymuria of clinical importance?

This question is difficult to answer conclusively. Urinary excretion of enzymes such as NAG is, however, a sensitive indicator of other types of renal injury [26,27]. Increased NAG excretion is seen in patients with poorly controlled diabetes mellitus [28], hypertension [29], acute tubular necrosis, renal transplant rejection, pyelonephritis with normal GFR, and after nephrotoxic drugs such as aminoglucosides [26]. Urinary excretion of tubular enzymes is also increased in patients with reduced GFR, irrespective of the aetiology of the kidney disease [30]. Interestingly, urinary excretion of NAG and ALP was not increased when iopentol was given to patients with severe renal failure [30]. Moreover, neither renal enzymes nor brush border antigens were increased after iohexol administration in patients with moderate-to-severe chronic renal failure (BG Danielson, personal communication). The reason for this phenomenon is not clear. It may be that the tubular cells are 'exhausted', and that chronic reduced renal function may have led to a reduction in the parenchymatous enzymatic pool, making the cells insensitive to CM.

18

However, in one study of patients with moderate renal failure (mean creatinine clearance 35 ml/min), urinary excretion of NAG, AAP and gamma-glutamyl-transferase (GGT) was increased by the HOCM diatrizoate, but not by the LOCM iopamidol [31].

Obviously, we cannot rely upon urinary enzyme excretion as being a marker for CM-induced nephrotoxicity in patients with moderate-to-severe reduced GFR (< 50 ml/min). This does not imply that CM are not tubulotoxic in such patients. CM exert a low, but significant chemotoxicity on several enzyme systems [32]. Patients with impaired GFR often suffer renal ischaemia. In suspensions of rabbit proximal tubule segments incubated with CM, at comparable concentrations to those reached in the tubules of patients, CM produced significant alterations in the intracellular content of potassium, calcium and ATP and uncoupled respiration effects, most pronounced after hypoxic cell injury [3]. CM, therefore, potentiated ischaemic cell injury in this model. Renal ischaemia frequently occurs in patients with impairment of GFR and may add to CM-induced tubular damage in patients with renal failure. Uni-nephrectomized, salt-depleted rats injected with indomethacin developed acute renal failure after administration of iothalamate. This type of acute renal failure was characterized by necrosis of the thick ascending limb of Henle [33].

CONCLUSIONS

In conclusion, as renal tubular changes are obviously important in the pathogenesis of CM-induced renal failure, the investigation of other discrete tubular function parameters such as tubular transport function could be important. Lithium clearance, for instance, which is a parameter for proximal tubular transport, has not been investigated after CM administration. The renal extraction of para-amino-hippuric acid is reduced by CM [34–35].

In the future, investigations of CM should include other parameters for proximal tubular transport and function. One should also investigate detailed renal glomerular and tubular function parameters in high-risk patients, such as patients with moderate-to-severe renal function impairment. Furthermore, coadminis-tration of CM and other drugs should be studied. Such investigations should include both nephrotoxic drugs and renal protective drugs.

REFERENCES

1. Olsson B, Aulie Å, Sveen K, Andrew E. Human pharmaco-kinetics of iohexol. A new nonionic contrast medium. Invest Radiol. 1983; 18: 177–182.
2. Waaler A, Svaland M, Fauchald P, Jakobsen JÅ, Kolmanns-kog F, Berg KJ. Elimination of iohexol, a low osmolar nonionic contrast medium, by hemodialysis in patients with chronic renal failure. Nephron. 1990; 56: 81–85.
3. Humes HD, Hunt DA, White MD. Direct toxic effect of the radiocontrast agent diatrizoate on renal proximal tubule cells. Am J Physiol. 1987; 252: F246–F255.
4. Andersen K-J, Vik H. Use of epithelial cell lines for testing cellular toxicity. In: Bianci C, Bocci V, Carone FA, Rabkin R, eds. Kidney, proteins and drugs: An update. Contrib Nephrol. Basel: Karger. 1993; 101: 227–234.
5. Nordby A, Halgunset J, Haugen OA. Effects of radiographic contrast media on monolayer cell cultures. Invest Radiol. 1986; 21: 234–239.
6. Svaland M, Kolmannskog F, Lillevold PE, Nordal KP, Ressem L, Berg KJ. Pharmacokinetics of iopentol in patients with chronic renal failure. Acta Radiol. 1992; 33: 482–484.
7. Barrett BJ, Parfrey PS, Vavasour HM, et al. Contrast nephropathy in patients with impaired renal function: High versus low osmolar media. Kidney Int. 1992; 41: 1274–1279.
8. Davidson CJ, Hlatky M, Morris KG, et al. Cardiovascular and renal toxicity of a nonionic radiographic contrast agent after cardiac catheterization. A prospective trial. Ann Intern Med. 1989; 110: 119–124.
9. Lautin EM, Freeman NJ, Schoenfeld AH, et al. Radiocontrast-associated renal dysfunction: Incidence and risk factors. Am J Roentgenol. 1991; 157: 49–58.
10. Schwab SJ, Hlatky MA, Pieper KS, et al. Contrast nephrotoxicity: A randomized controlled trial of a nonionic and an ionic radiographic contrast agent. N Engl J Med. 1989; 320: 149–153.
11. Manske CL, Sprafka JM, Strony JT, Wang Y. Contrast nephropathy in azotemic diabetic patients undergoing coronary angiography. Am J Med. 1990; 89: 615–620.
12. Berg KJ, Jakobsen JÅ. Nephrotoxicity related to contrast media. In: Enge I, Edgren J, eds. Patient safety and adverse events in contrast medium examinations. Amsterdam: Elsevier Science Publishers. 1989: 111–120.
13. Byrd L, Sherman RL. Radiocontrast-induced acute renal failure: A clinical and pathophysiological review. Medicine (Baltimore). 1979; 58: 270–279.
14. Campbell DR, Flemming DK, Mason W, et al. A comparative study of the nephrotoxicity of iohexol, iopamidol and ioxaglate in peripheral angiography. J Can Assoc Radiol. 1990; 41: 133–137.
15. Barrett BJ, Carlisle EJ. A meta-analysis of the relative nephrotoxicity of high and low-osmolality iodinated contrast media. Radiology. In press.

16. Kunin CM, Chesney RW, Craig WA, England AC, DeAngelis C. Enzymuria as a marker of renal injury and disease: Studies of N-acetyl-β-glucosaminidase in the general population and in patients with renal disease. Pediatrics. 1978; 62: 751–760.

17. Scherberich JE, Rautschka E, Fischer A, Kollath J, Riemann EH. Tubular histuria: Clinical evaluation of the different nephrotoxic potential of X-ray contrast media. In: Bianci C, Bocci V, Carone FA, Rabkin R, eds. Kidney, proteins and drugs. Contrib Nephrol. Basel:Karger. 1990; 83: 229–236.

18. Jakobsen JÅ, Berg KJ, Waaler A, Andrew E. Renal effects of the non-ionic contrast medium iopentol after intravenous injection in healthy volunteers. Acta Radiol. 1990; 31: 87–91.

19. Jakobsen JÅ, Nossen JØ, Jørgensen NP, Berg KJ. Renal tubular effects of diuretics and X-ray contrast media: A comparative study of equi-molar doses in healthy volunteers. Invest Radiol. In press.

20. Powell CJ, Holtz, E, Bridges JW. Nephrotoxicity of non-ionic and iso-osmotic dimeric radiological contrast media. J Pathol. 1985; 146: 276.

21. Hofmeister R, Bhargava AS, Günzel P. The use of urinary N-acetyl-β-D-glucosaminidase (NAG) for the detection of contrast-media-induced "osmotic nephrosis" in rats. Toxicol Lett. 1990; 50: 9–15.

22. Moureau J-F, Droz D, Sabto J, et al. Osmotic nephrosis induced by water- soluble triiodinated contrast media in man. Radiology. 1975; 115: 329–336.

23. Jakobsen JÅ, Lundby B, Kristoffersen DT, Borch KW, Hald JK, Berg KJ. Evaluation of renal function with delayed CT after injection of nonionic monomeric and dimeric contrast media in healthy volunteers. Radiology. 1992; 182: 419–424.

24. Kløw NE, Levorstad K, Berg KJ, et al. Iodixanol in cardioangiography in patients with coronary artery disease. Tolerability, cardiac and renal effects. Acta Radiol. 1993; 34: 72–77.

25. Pugh ND, Sissons GRJ, Ruttley MST, Berg KJ, Nossen JØ, Eide H. Iodixanol in femoral arteriography (phase III): A comparative double-blind parallel trial between iodixanol and iopromide. Clin Radiol. 1993; 47: 96–99.

26. Price RG. Urinary enzymes, nephrotoxicity and renal disease. Toxicology. 1982; 23: 99–134.

27. Wellwood JM, Ellis BG, Price RG, Hammond K, Thompson AR, Jones NF. Urinary N-acetyl-β-D-glucosaminidase activities in patients with renal disease. Br Med J. 1975; 3: 408–411.

28. Whiting PH, Ross IS, Borthwick LJ. N-acetyl-D-glucosaminidase levels and diabetic microangiopathy. Clin Chim Acta. 1979; 97: 191–195.

29. Alderman MH, Melcher L, Drayer DE, Reidenberg MM. Increased excretion of N-acetyl-β-glucosaminidase in essential hypertension and its decline with antihypertensive therapy. N Engl J Med. 1983; 309: 1213–1217.

30. Berg KJ, Kolmannskog F, Lillevold PE, et al. Iopentol in patients with renal failure: its effects on renal function and its use as a glomerular filtration parameter. Scand J Clin Lab Invest. 1992; 52: 27–33.

31. Cavaliere G, Arrigo G, D'Amico G, et al. Tubular nephrotoxicity after intravenous urography with ionic high-osmolal and nonionic low-osmolal contrast media in patients with chronic renal insufficiency. Nephron. 1987; 46: 128–133.

32. Dawson P. Chemotoxicity of contrast media and clinical adverse effects: A review. Invest Radiol. 1985; 20 (suppl): S84–S91.

33. Heyman SN, Brezis M, Greenfeld Z, Rosen S. Protective role of furosemide and saline in radiocontrast-induced acute renal failure in the rat. Am J Kidney Dis. 1989; 14: 377–385.

34. DiBona GF. Effect of anionic and nonionic contrast media on renal extraction of para-aminohippurate in the dog. Proc Soc Exp Biol Med. 1978; 157: 453–455.

35. Talner LB, Davidson AJ. Effect of contrast media on renal extraction of PAH. Invest Radiol. 1968; 3: 301–309.

This paper was first published in *Advances in X-Ray Contrast*. 1993;1:10–18.

P. Dawson and W. Clauss, (eds.), Advances in X-Ray Contrast: Collected Papers. 20–28
© *1998 Kluwer Academic Publishers.*

The role of contrast agents in thromboembolic phenomena in clinical angiography

Peter Dawson, PhD, MRCP, FRCR
Royal Postgraduate Medical School, Hammersmith Hospital, Du Cane Road, London, W12 ONN, UK

The risk of embolization of the patient has long been recognized as a potential hazard in both diagnostic and interventional angiography. The dangers are, naturally, particularly serious in the coronary and cerebral vessels. Embolization may be caused by fibres from swabs [1,2], air from lines, debris found routinely in guidewires and catheters, even when new [3], particulate contaminants routinely found in contrast agent solutions [3], cholesterol [4], atheromatous material [5] and thrombus. The last is of special interest because it is here that the angiographer can most reliably minimize the risk by use of good technique, and it is on this area of thromboembolism, therefore, that this article will concentrate.

The formation of blood clots in catheter lines and syringes, with the attendant risk of injection into the patient, has been of particular concern for angiographers for many years [6]. The materials of catheters and syringes are relevant to the phenomenon but most angiographers would agree that the problem is more one of technique than of materials. The role of the contrast agents, so vital to these procedures, has, until recently, attracted very little attention, though there has long been a vague perception in the angiographic community that contrast agents possess anticoagulant properties which, up to a point, are beneficial to the procedure [7]. Recent reports in the literature [8] that blood-contaminated syringes containing non-ionic, as opposed to either conventional or low osmolality ionic agents, are more likely to generate clots have generated considerable interest. Indeed, in the United States it has been the subject of perhaps the most heated controversy in the history of angiography. It is certain that the medicolegal background against which US angiographers and cardiologists have to work has been an important influence in its development. However, the interactions between contrast agents and blood in vitro in syringes and catheters, and in vivo, must be considered seriously if a proper understanding of the problem is to be achieved.

HISTORICAL

A variety of studies over the years [9–12] has established that conventional ionic contrast media have anticoagulant effects, due in part to the inhibition of fibrin polymerization, and perhaps, in part, to the binding of calcium. In addition, it has been observed that, in vitro, contrast agents have dose-dependent inhibitory effects on normal platelet aggregation [11,12]. In 1957, McAfee reported that an occasional haemorrhagic tendency could, in fact, be observed in association with contrast media and that this was indeed their most important haematological effect [13]. The more recent work [11,12] has confirmed earlier observations and established that there are quantitative, but not qualitative, differences between the effects of different types of agents in this regard.

Clinical thromboembolic events, particularly in coronary angiography, began to attract special attention in the 1970s. Adams et al [14] reported a frequency of 0.5% for myocardial infarct and 0.23% for cerebral embolization in a study of some 47 000 patients undergoing coronary angiography with the (then) conventional ionic agents. Surprisingly, the frequency seemed to be related to the route of vascular access, with the brachial approach yielding complication rates of 0.22% and 0.03%, respectively, and the femoral approach yielding rates of 1.01% and 0.43%, respectively. At about the same time, Takaro [15] studied 34 cases of acute coronary occlusion. He found fresh clots at autopsy in all 22 fatal cases, and at the time of emergency surgery in four non-fatal cases. The angiographic procedures had all been performed by means of a femoral approach and involved catheter changes. The authors suggested the mechanism

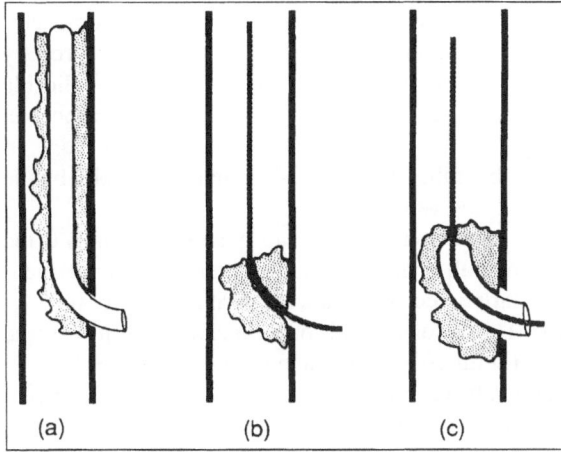

Figure 1 Thrombus builds up on the *outside* of a catheter. This is stripped off when the catheter is removed. It may plug the hole, embolize the distal vessel or be picked up by a replacement catheter (c), or carried to some new site.

illustrated in Figure 1 to explain this phenomenon.

In spite of the clear fact emerging from this study that most of the events observed were truly due to thrombus formation in or on the catheter, it is important to understand that it is difficult clinically to separate such phenomena as dissection, distal (non-embolic) spontaneous thrombotic closure, cardiac arrhythmia without vessel occlusion, dislodging of atheromatous plaque, etc. It may very well be that few acute events ascribed to thromboembolism actually are in reality.

In a subsequent study, Takaro et al [16] reported 66 deaths associated with coronary angiography in 3044 patients. The coronary occlusions accounting for 35 fatal cases were due to thromboembolism in 26, intraplaque haemorrhage in two and dissection in one. In the remaining six cases, no pathological support was found for the angiographic diagnosis of acute vessel occlusion. These facts support, to some extent, the caveat mentioned above. There were four cerebral emboli, and at least 30 of the 66 deaths in approximately 1% of all patients were thought attributable to thromboembolic complications – a frequency close to that found by Adams et al among their patients undergoing coronary angiography via the femoral approach.

Other workers in the 1970s found similar or higher frequencies of thromboembolism. Green et al [17] observed two myocardial infarcts and two peripheral emboli in 445 patients. This is, once again, a frequency of about 1%. Six myocardial infarcts were observed by De la Torre [18] among 139 procedures, a startling 4% frequency.

Judkins and Gander [18] examined the data of Adams et al [14] and also considered other data and, in a much-cited paper, emphasized the major role of personnel experience. They argued that the femoral approach was preferred by less-skilled personnel and was not usually performed with systemic heparinization of the patient. On the other hand, with the brachial approach, the risk to the brachial artery was usually uppermost in the operator's mind and systemic heparinization was commonly employed. In fact, Judkins and Gander noted that the incidence of thromboembolism in medical centres performing fewer than 100 coronary arteriograms per year was some 10 times higher than centres performing more than 400 per year. They stressed the need to reduce the procedure time, to use trained personnel or good supervision, to use less thrombogenic materials (a counsel of perfection) and to institute systemic heparinization. As regards the last, it is important to understand that, important as they thought its place might be, they considered it less important than technique and experience.

In all this, it should also be noted that any role of contrast media, positive or negative, did not enter into their considerations.

Systemic administration of heparin was not universally accepted in the 1970s. Some investigators supported the idea [20,21] but others argued that aspirin was more effective [22]. Still others pointed out the risk of fatal haemorrhage in a patient given anticoagulants in whom catheter-induced arterial perforation occurred [23].

Interest in the thrombogenicity of catheter and guidewire materials grew and a number of studies were reported which concluded that the growth rate of thrombus on a catheter is:

(a) non-linear and related to the surface material and finish of the catheter
(b) greatest for Teflon-coated materials
(c) less for polyurethane materials and

22

(d) least for polyethylene materials.

The porous surface of Teflon and the relative surface irregularity of polyurethane in most preparations, which may be demonstrated by electron microscopy, both play a part in the basis for these observations. Catheter and guidewire thrombogenicity varies from manufacturer to manufacturer, however, as much as it does from material to material. Heparin-coated or heparin-impregnated catheters are expensive and do not have clearly established advantages. Teflon-coated guidewires are very thrombogenic but provide reduced frictional resistance, especially when they are used with polyurethane catheters. Stainless steel wires may be polished or unpolished, the former being, predictably, less thrombogenic that the latter. Wires should remain in the circulation for the shortest possible time and should be wiped with a wet rather than a dry swab after each use, as the latter may cause trapping of cotton fibres in the wire, with subsequent fibre embolization.

Some authors in this period recommended that contrast media should be used as flushing agents because of their anticoagulant properties [24]. It was clear that, at this stage of the evolution of angiography, contrast agents were seen as part of the solution and not as part of the problem. In a follow-up survey performed at the end of the 1970s, Adams and Abrams [25] reviewed a very large number of coronary angiograms and noted a pronounced decrease, since the publication of Judkins and Gander [19], in fatal complications associated with the femoral approach, to a level close to that found with the brachial approach. The authors ascribed this not so much to a wider use of systemic heparinization of patients, though there was some evidence that this occurred, but to the greater expertise and care on the part of practitioners. Davis et al [26] also presented data at about the same time, obtained on an entirely different basis from that of Adams and Abrams but, nevertheless, supportive of the thesis that meticulous technique, rather than the use of systemic heparinization, was the basis for the improvement. More recently, Kennedy et al [27] reported a 1.8% overall complication rate in some 54 000 patients undergoing coronary angiography. Their conclusion was that as many as 60% of the fatal complications were thromboembolic and that the overall frequency of thromboembolism was approximately 0.25%.

Although most of the documentation of thromboembolic phenomena was in the field of coronary angiography, the general principles seem applicable to cerebral and other angiographies. It is reasonable, therefore, to summarize the state of knowledge of thromboembolic complications in angiography by the mid-1980s, as follows:

(a) Thromboembolism contributes significantly to the overall complication rate in angiography, particularly in coronary and cerebral angiography.
(b) The use of systemic heparinization may have some useful effect in reducing frequency.
(c) Catheter and guidewire materials play a significant role in causation but, though some materials are better than others, the end product produced by different manufacturers is variable.
(d) The contrast agents have anticoagulant properties and antiplatelet aggregation properties and may be playing some part in reducing the frequency of problems.
(e) Most importantly, the risk appears to be related to experience and technique. For example, for experienced angiographers, the frequency of thromboembolic complications in coronary angiography is approximately 0.25%. In less experienced hands, historically, the frequency would appear to be an order of magnitude higher.

CONTRAST AGENTS

Attention was refocused on thromboembolic phenomena in angiography in 1987, but with a different emphasis from previously, following two publications [8,29]. During the previous few years, a new generation of non-ionic and ionic, dimeric, low osmolality contrast agents had been introduced [28]. These were better tolerated by patients and, it was suspected, possibly safer with regard to the frequency of major anaphylactoid reactions. This has now been firmly established by large-scale clinical trials. Within the limits inevitably imposed by the greater cost, these low osmolar agents established for themselves a major role in European radiological practice and, since 1985, a growing role in US practice. The two publications referred to above changed the atmosphere where contrast agents were concerned.

Raininko and Ylinen [29] observed that, in vitro,

non-ionic contrast media may engender a disordered red blood cell aggregation phenomenon similar to that produced by dextrose solutions [30]. They reasoned that if such aggregates were to form in syringes or catheters they might behave as emboli. At about the same time, Robertson [8] published his account of clot formation in blood-contaminated syringes. This phenomenon occurred more frequently with non-ionic than with ionic contrast agents, albeit after a long incubation period of several tens of minutes. Both articles seemed to indicate that the new agents might actually pose a previously unsuspected danger of thromboembolism in angiography, for all that they were safer in other regards. This was the first time that any contrast agent had been held to play any role other than a helpful, inhibitory one in thromboembolic accidents in angiography. Important associated clinical, medicolegal and commercial aspects served, on occasion, to generate more heat than light in this debate.

While these observations cannot be ignored, it should be noted that they were based entirely on observations in vitro and, in the latter case, were somewhat removed from the real world of clinical angiography, in that blood was allowed to remain contaminated by contrast agents in syringes for excessively long periods, in order to obtain clots in a few cases. It should also be noted that there had been no impression of any increase in the incidence of thromboembolic phenomena in clinical angiography in Europe or the United States during the several years of use of non-ionic contrast agents up to that point. However, in 1988, in the wake of Robertson's paper, Grollman and colleagues [31] published a report of three cases of coronary embolization associated with the use of non-ionic contrast agents, in a survey of 1380 coronary angioplasty procedures. This actually represented a frequency of complications lower than the previously reported figure of 0.25% for coronary studies performed by experienced angiographers with conventional contrast agents but, nevertheless, generated considerable anxiety. Bashore and colleagues [32] observed two cases of coronary occlusion in 2650 procedures performed with an ionic agent, and five cases in 3313 procedures with a non-ionic agent. There was no statistically significant difference between these two, and the absolute frequency of complications with both media was, again, well below the accepted rate of 0.25%. However, once again, this seemed to fuel the controversy.

A number of alleged cases were reported to the FDA Spontaneous Reporting System and, in 1988, 16 cases of allegedly associated complications with iopamidol, 14 with iohexol (both non-ionics) and 6 with ioxaglate (ionic, low osmolality) had been filed [33]. These data served only to confuse the issue in that, given the less frequent use of ioxaglate compared with the non-ionic agents, they appeared to suggest a higher rate of thromboembolism with the former. Reporting systems usually under-report events but, at times of high anxiety about a drug or agent, some systems may actually over-report and a proportion of these may be of a medicolegally defensive character. The controversy did, however, stimulate new work and some new thoughts on older work in this field, which will now be reviewed.

RED CELL AGGREGATION

It has been known for some years that, even in iso-osmolar concentrations, contrast agents engender changes in red blood cell morphology and inhibit Rouleaux (ordered aggregates) formation. Non-ionic agents were found to inhibit Rouleaux formation less vigorously. There is evidence that ioxaglate actually stimulates Rouleaux formation to some extent, but none of this is of great importance, since Rouleaux are clinically unimportant entities. Raininko and Ylinen observed the different phenomenon reproduced in Figure 2. In Figure 2(a), ionic contrast material in contact with blood on a microscope slide produces the occasional small disordered aggregate but, as shown in Figure 2(b), the non-ionic agent produces much larger, more numerous aggregates. The same phenomenon may be seen in syringes (Figure 3) and has been confused with true clot formation [34]. Dawson [35], Zucker and Mauss [36] showed that even small changes in the chemical environment can result in complete disruption of such aggregates and Aspelin et al [37] and Grabowski [38] have shown that only a very low shear is needed to disrupt them. The opinion of all these authors is that, although interesting, the phenomenon is almost certainly of no clinical importance.

24

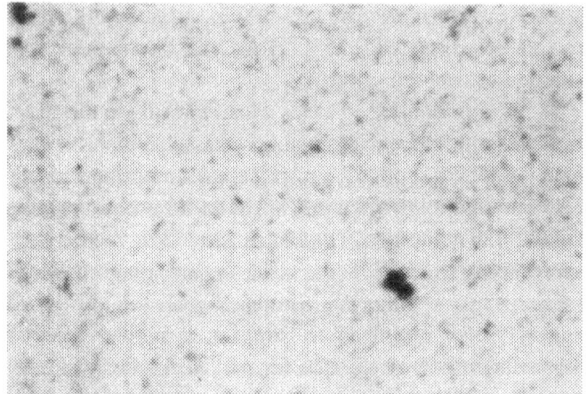

(a)

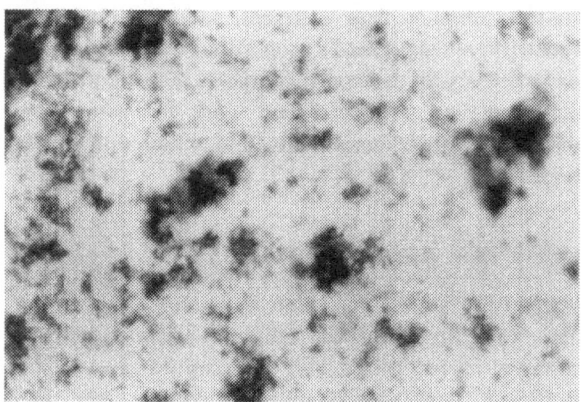

(b)

Figure 2 (a) Formation on a microscope slide of small, scattered, red cell aggregates in the presence of an ionic contrast agent.
(b) Formation of many large disordered red cell aggregates in the presence of non-ionic agents.

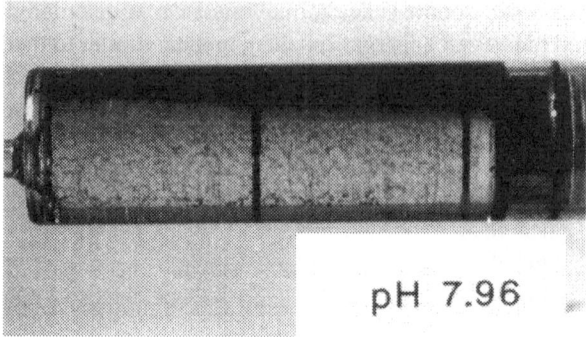

pH 7.96

Figure 3 Red cell aggregates forming in a syringe of non-ionic contrast agent contaminated by blood.

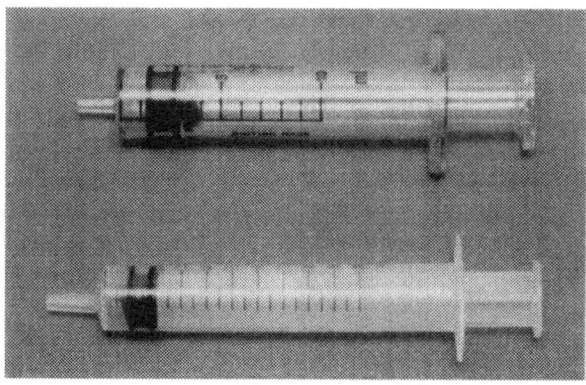

Figure 4 Two syringes in widespread use made of different plastics with different thrombogenicities.

THROMBUS FORMATION

Dawson [35] attempted to reproduce Robertson's observations on true clot formation by contaminating a large number of syringes containing various contrast media with blood and leaving them to stand for long periods. Clots formed only in the presence of non-ionic agents and then only in about one syringe in 50, in circumstances that represented a marked deviation from good angiographic technique. Furthermore, clots formed in this study in the syringes made of one kind of material but not of the other (Figure 4).

Clots did not form in any syringe in which a deliberate effort was made to mix the blood and contrast agent thoroughly. Further studies were done in which fibrinopeptide A generation was measured to determine the contact-activating propensity of the different syringe materials [35]. These studies confirm that glass is a powerful activator but that not all plastic syringe materials are equal in this regard (though all are better than glass). Polypropylene is the less active of the two plastic materials studied.

The conclusions of this work were that, while contrast agents inhibit clot formation to different degrees, contact activation by the syringe and catheter surfaces is also very important. If blood contaminating the syringes is not deliberately mixed it separates from the contrast agent, floats on top of it, and is forced into contact with the syringe material and undergoes contact activation. In these circumstances, the poor mixing and the resultant decreased contact between contrast material and blood tends to minimize the clotting-inhibitory effect of the contrast agent.

Figures 5 and 6 show the effects of increasing concentrations of contrast agents on the thrombin time and on collagen-activated platelet aggregation obtained by Dawson et al [11]. Similar data have been obtained by Stormorken et al [12]. Clearly, the non-ionic agents have lesser effects than the ionic agents (high or low osmolality). All contrast media have anticoagulant properties, but non-ionic agents with their relatively inert behaviour in many biological systems [39] have the least marked effects.

CATHETER FLUSHING TECHNIQUE

Adequate and frequent catheter flushing is an important aspect of good angiographic technique. Dahlborn et al [40] suggested that clotting could be prevented by flushing the catheter with a simple saline solution at a minimum rate of 4 ml per minute. Some angiographers flush continuously, with or without intermittent pulses of heparinized saline. However, with some catheters the pressure of flushing is more important than the nature of the flush fluid (Figure 7). Figure 7(a) shows a Pigtail catheter placed under water and flushed gently, but at a rate of more than 4 ml per minute, with a blue dye. Only the proximal side-holes are flushed. Figure 7(b) shows that the other side-holes are flushed at a higher injection rate and pressure; it requires a still greater pressure to flush the end-hole, as is shown in Figure 7(c). Continuous, low-pressure flushing is inadequate with catheters of this type. Clots may form in the distal portion, only to be ejected after a rapid, high-pressure injection of contrast material, with sometimes disastrous results.

It has been said that the smaller the catheter used, the less chance of thrombus formation; therefore, 5F catheters should be safer than 7F catheters of the same material. This statement may be broadly true but, in the case of catheters with multiple side-holes, like that illustrated in Figure 7, it is not entirely true because of the aforementioned problem of clot formation in the tip

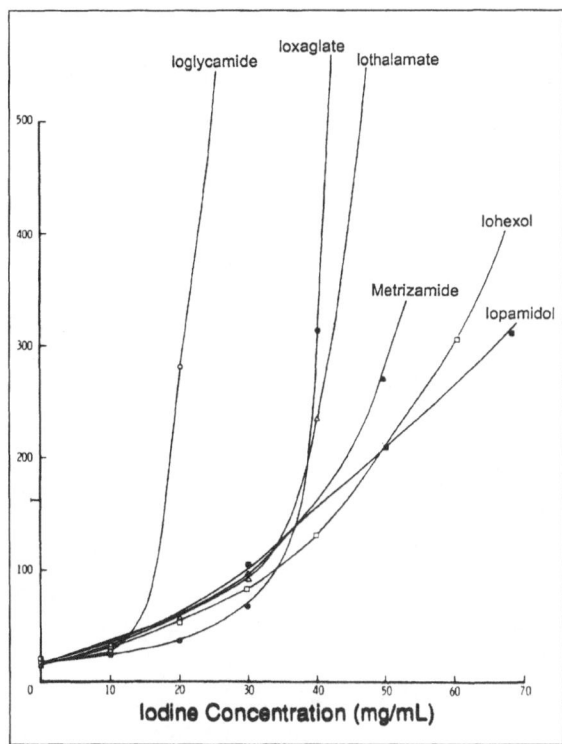

Figure 5 Prolongation of the thrombin time by contrast agents. The effect is dose-dependent and contrast agent type-dependent. The non-ionic agents have the least effect but still are anticoagulant.

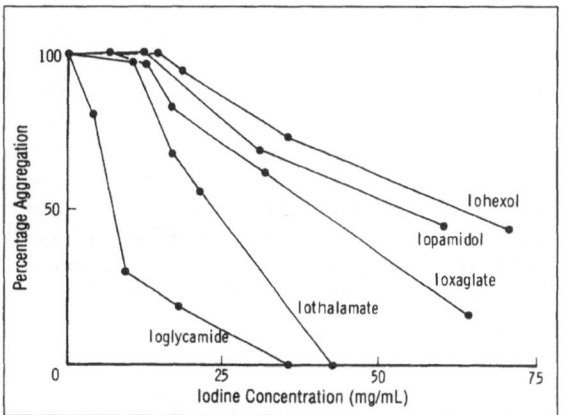

Figure 6 Inhibition of platelet aggregation by contrast agents. The effect is dose-dependent and contrast agent type-dependent. The non-ionic agents have the least effect, but, nevertheless, a positive one.

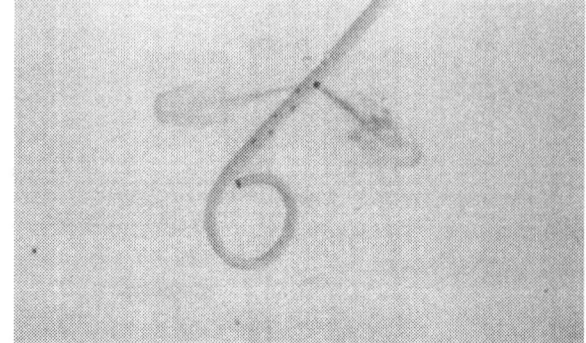

(a)

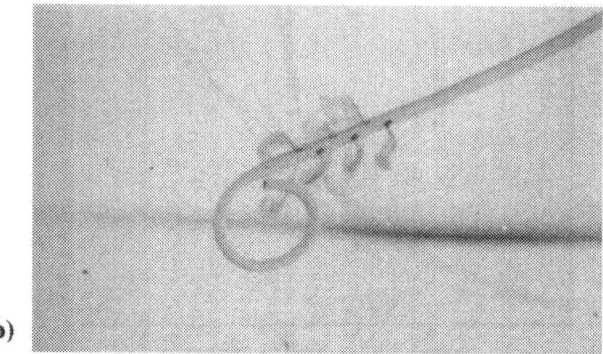

(b)

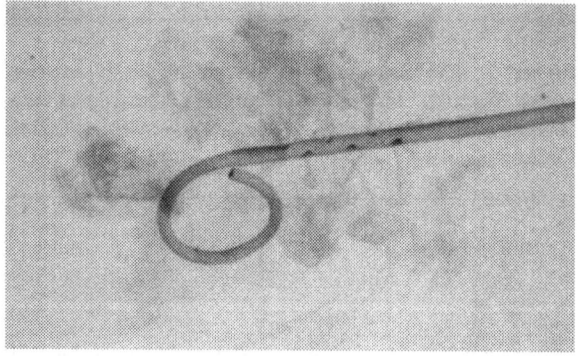

(c)

Figure 7 The problem of flushing a pigtail catheter. Gentle or continuous flushing clears the proximal side-holes only (a); harder flushing clears all the side-holes (b); a very hard flush is essential to be sure of flushing the end-hole (c).

of the catheter beyond the side-holes. The pressure required to flush distal side-holes and end-holes is higher with the smaller catheters than with the larger ones. We have seen the formation of thrombi in the distal portion of 3F and 5F polyurethane Pigtail

catheters within minutes of their insertion, despite frequent flushing with heparinized saline and the use of an ionic contrast medium.

HEPARINIZATION

In a recent study, Hwang and colleagues [41] observed that clots could be introduced in glass syringes (glass is a very effective contact activator and should be avoided in clinical angiography when possible), in mixtures of blood and non-ionic contrast agents and in blood–saline mixtures that are incubated over a period of approximately 30 minutes. Clots were never present when the blood was taken from a patient given 2000 IU of heparin systemically. However, the amount of heparin required to achieve a therapeutic level in a given patient varies enormously, making the determination of the given dose very difficult [42].

A recent study by Miller [43], of members and Fellows of the Society of Cardiovascular and Interventional Radiology in the United States, revealed considerable differences across the US in the mean heparin dose used in flush solutions for angiography and, especially, in the frequency with which angiographers use heparin with non-ionic contrast agents. The routine use of systemic heparinization in patients undergoing diagnostic angiography was rare. Heparin concentrations in flush solutions varied from 0-12 000 IU/L. The authors concluded that the survey revealed no standard of practice, that the heparin dose must be adjusted for each procedure and each patient, and that further, large-scale studies were needed to determine the merits of systemic heparinization.

It is also important not to lose sight of the fact that an effort, clinically equivalent to very bad angiographic technique, has to be put into creating clots, as evidenced by the use of glass syringes and by the 30-minute incubation periods used by Hwang and colleagues [39]. Furthermore, as indicated earlier, available clinical data do not substantiate the existence of a problem. We may quote Bettmann [44] "experience suggests that these differences in clotting (between ionic and non-ionic agents) are far less important than the combination of training, judgement and careful adherence to optimal technique".

SUMMARY

Thromboembolism during angiography occurs with a frequency of about 0.25% in good hands. Recent data have shown that the ionic agents are less effective anticoagulants than the ionic agents. They are, nevertheless, anticoagulants and should still be seen as part of the solution and not as part of the problem, as regards thromboembolic events in clinical angiography. There is no evidence that they have any prothrombotic potential.

Other important factors are syringe and catheter materials, catheter size, duration of the procedure and flushing technique. Flushing should be frequent and vigorous. The individual patient's clotting status may clearly play a part, but the role of systemic heparinization is not established and cannot be relied upon, since the heparin 'requirements' of individuals vary widely. However, experience and technique are the most important factors of all, and we can do no better in the 1990s than to remember the injunction of Judkins and Gander [19] in the 1970s: "it is the fine points of technique that make the difference in complication rates... (coronary) arteriography is unforgiving. If near perfection is to be achieved, more than heparinization is needed. It is the little things that count"

REFERENCES

1. Adams DF, Olin TB, Kose KJ. Cotton fiber embolization during angiography: a clinical and experimental study. Radiology. 1965; 165: 678–681.
2. Kay JM, Wilkins RA. Cotton fiber embolization during angiography. Clin Radiol. 1969; 20: 410–413.
3. Winding O. Contaminants in contrast media and catheters. In: Ansell G, Wilkins RA, eds. Complications in diagnostic imaging, 2nd edn. Oxford: Blackwell Scientific; 1987.
4. Ramirez G, O'Neill WM, Lambert R, Bloom MA. Cholesterol embolization: a complication of angiography. Arch Intern Med 1978; 138: 1430–1432.
5. Perdue GD, Smith RB. Atheromatous microemboli. Ann Surg. 1969; 169: 954–959.
6. Hessel SJ, Adams DE, Abrams ML. Complications of angiography. Radiology. 1981;138: 273–281.
7. Katzen BT. Interventional and diagnostic therapeutic procedures. New York: Springer-Verlag; 1980: 2–3.
8. Robertson HJF. Blood clot formation in angiographic syringes containing nonionic contrast media. Radiology. 1987; 162: 621–622.
9. Bernstein EF, Gans H. Anticoagulant activities of angiographic contrast media. Invest Radiol. 1988; 1: 162–164.
10. Verbeley K, Kutt M, Torack RM, McDonell F. The effects of radiopaque contrast media on the structures and solubility of the fibrin clot. Blood. 1969; 33: 468–478.
11. Dawson P, Hewitt MR, Mackie IJ, Machin SJ, Amin S, Bradshaw A. Contrast, coagulation and fibrinolysis. Invest Radiol. 1986; 21: 248–252.
12. Stormorken H, Skalpe IO, Testart MC. Effect of various contrast media on coagulation, fibrinolysis and platelet function: an in vitro and in vivo study. Invest Radiol. 1986; 21: 348–354.
13. McAfee JG. Survey of complications of abdominal arteriography. Radiology. 1957; 68: 825-827.
14. Adams DF, Fraser DB, Abrams HC. The complication of coronary arteriography. Circulation. 1973; 48: 609–612.
15. Takaro T, Pifarre R, Wuerflein RD. Acute coronary occlusion following coronary arteriography: mechanism and surgical arteriography: mechanism and surgical relief. Surgery. 1972; 72: 1018–1029.
16. Takaro T, Hultgren HH, Littman D. An analysis of deaths occurring in association with coronary arteriography. Am Heart J. 1973; 86: 587–597.
17. Green GS, McKinnon CM, Roech J. Complications of selective percutaneous coronary arteriography and their prevention: a review of 445 consecutive examinations. Circulation. 1972; 45: 552–557.
18. De la Torre J, Jacobs D, Aleman J, Anderson GA. Embolic coronary artery occlusion in percutaneous transfemoral coronary arteriography. Am Heart J. 1973; 86: 467–473.
19. Judkins MP, Gander MP. Complications of coronary arteriography. Circulation. 1974; 49: 599–602.
20. Walker WT, Murdall SL, Broderick D. Systemic heparinization for femoral percutaneous coronary arteriography. N Engl J Med. 1973; 288: 826–828.
21. Eyer KM. Complications of transfemoral coronary arteriography and their prevention using heparin. Am Heart J. 1973; 86: 428.
22. Sarlman EW. The limitations of heparin therapy after arterial reconstruction. Surgery. 1965; 57: 131–138.
23. Storstein O, Nitter-Hange S, Enger I. Thromboembolic complications in coronary angiography. Acta Radiol [Diagn] (Stockh). 1977; 18: 555–560.
24. Hawkins IF, Herbert L. Contrast material used as catheter flushing agent: a method to reduce clot formation during angiography. Radiology. 1974; 110: 351–352.
25. Adams DF, Abrams HL. Complications of coronary arteriography: a follow up report. Cardiovasc Radiol. 1979; 2: 89–92.
26. Davis K, Kennedy JW, Kemp MG, Judkins MP. Complications of coronary arteriography from the collaborative study of coronary artery surgery (CASS). Circulation. 1979; 59: 1105–1112.
27. Kennedy JW. Complications associated with cardiac catheterization and angiography. Cathet Cardiovasc Diagn. 1982; 8: 5–9.
28. Dawson P, Pitfield J, Grainger RG. The low osmolality contrast agents: a simple guide. Clin Radiol. 1983; 34: 221–226.

28. Dawson P, Pitfield J, Grainger RG. The low osmolality contrast agents: a simple guide. Clin Radiol. 1983; 34: 221–226.

29. Raininko R, Ylinen SL. Effect of ionic and non-ionic contrast media on aggregation of red blood cells in vitro. Acta Radiol [Diagn] (Stockh). 1987; 28: 87–92.

30. Wilson H. Aqueous dextrose solutions: a hazard in transfusion. Am J Clin Pathol. 1950; 20: 667–669.

31. Grollman JH, Liu CK, Astone RA, Lurie MD. Thromboembolic complications in coronary angiography associated with the use of non-ionic contrast medium. Cathet Cardiovasc Diagn. 1988; 14: 159–164.

32. Bashore TM, Davidson CK, Mark DB. Iopamidol use in the cardiac catheterization laboratory: a retrospective analysis of 3,313 patients. Cardio 1988; 5(3, pt 2): 60–100.

33. Robertson HJF. Non-ionic contrast media in radiology: procedural considerations. Invest Radiol. 1988; 23 (suppl): S374–S377.

34. Rasuli P. Blood clot formation in angiographic syringes containing non-ionic contrast media. Radiology. 1987; 163: 621–622.

35. Dawson P. Non-ionic contrast agents and coagulation. Invest Radiol. 1988; 23: S310–S317.

36. Zucker MB, Mauss EA. Erythrocyte aggregation in iohexol and other nonionic media. Invest Radiol. 1988; 23 (suppl):S340–S345.

37. Aspelin P, Birk A, Almen T, Kieseweller H. Effect of iohexol on human erythrocytes. l. Changes of red cell morphology in vitro. Acta Radiol. 1980; 362 (suppl): 123–126.

38. Grabowski EF. Effects of contrast media on erythrocyte and platelet interactions with endothelial cell monoiayers exposed to flowing blood. Invest Radiol. 1988; 23: S351–S358.

39. Dawson P. Chemotoxicity of contrast agents and clinical adverse effects. Invest Radiol 985; 20: 584–591.

40. Dahlborn M, Cronestrand R, Klintmal G, Sundelins S. Blood inflow and coagulation in vascular catheters: comparison of the effects of polysaccharide solutions, saline and contrast medium. Acta Radiol [Diagn] (Stockh). 1980; 21: 715–720.

41. Hwang MH, Piao ZE, Murdock DK, Messmore HL, Giardina JJ, Scanlon PJ. Risk of thromboembolism during diagnostic and interventional cardiac procedures with non-ionic contrast media. Radiology. 1990; 174: 453–457.

42. Hattersley PG. Heparin anticoagulation. In: Koepke JA, ed. Laboratory haematology. New York: Churchill Livingstone; 1984: 789–818.

43. Miller DL. Heparin in angiography: current patterns of use. Radiology. 1989; 172: 1007–1011.

44. Bettman MA. Guidelines for use of low osmolality contrast agents. Radiology. 1989; 172: 901–903.

This paper was first published in *Advances in X-Ray Contrast*. 1993;1:32–40.

UPDATE

Since publication of this review two major issues, which are in some ways linked, have attracted some attention in the angiography and cardiology communities. These are the future clinical role of the non-ionic dimers [1] and the observation by Chronos and colleagues [2] that non-ionic monomeric agents are capable, at least in vitro, of platelet activation and degranulation. The non-ionic dimers have a number of merits. They are isotonic with body fluids and have even lower chemotoxicity than the non-ionic monomers. They might find useful applications in high dose interventional studies in CT, and in some organ-specific studies such as cardioangiography [1].

However, they share with non-ionic monomers the low chemotoxicity which leads to minimal anti-coagulant effect. As the author has argued in the review, such fears are exaggerated and good technique with low toxicity (non-ionic) contrast agents is the ideal policy. However, the story of the role of contrast agents in clot formation, thromboembolism, etc., has taken a new twist with the work of Chronos and colleagues [2]. Their observation of platelet degranulation by non-ionic, but not ionic, agents is interesting. The phenomenon was observed in vitro but they postulate that similar in-vivo effects might occur making non-ionics effectively prothrombotic. There are important questions about the validity of extending these in vitro results to the clinical situation but there is another question about the authors' suggestion that this is a non-ionic phenomenon. The author and colleagues [3] have recently shown that this property of non-ionic monomers is not shared by the non-ionic dimers.

This reinforces the view that the non-ionic dimers may become agents of choice for some angiographic procedures.

References

1. Dawson P. The non-ionic dimers – some theoretical and clinical considerations. European Radiology 1995; 5: S101–S106.

2. Chronos NAF, Goodall AH, Wilson DJ, Sigwart V, Buller NP. Profound platelet degranulation is an important side effect of some types of contrast media used in interventional cardiology. Circulation 1993; 88: 2044–2045.

3. Dawson P et al. Unpublished data.

P. Dawson and W. Clauss, (eds.), Advances in X-Ray Contrast: Collected Papers. 29–33
© 1998 Kluwer Academic Publishers.

Delayed reactions to intravenous injections of urographic contrast media

Peter Davies, FRCR, DM
Nottingham City Hospital, Hucknall Road, Nottingham, NG5 1PB, UK

INTRODUCTION

There are two fundamental aspects to the study of reactions, either acute or delayed:
1. Clinical observation of individual syndromes.
2. Data collection to estimate the prevalence.

These are, to some extent, in conflict, as accurate clinical observation of each reaction by a defined group of workers is not possible on the large numbers of patients from different hospitals required for statistical rigour.

A third point is how important are such reactions and how they are defined. There is again a conflict, as the better the definition of the endpoint, the smaller the numbers of patients to be studied.

If one considers death as the endpoint, one is still left with some uncertainties. Pendergrass et al [1] classified their cases into immediate, delayed and indeterminate. An immediate reaction was one occurring during the examination, while delayed reactions occurred more than an hour later. These authors refer only briefly to the delayed and intermediate cases because of the difficulty in establishing a direct cause-and-effect relationship with the injection. They also note two patients who died immediately before the injection of contrast medium from 'coronary attacks' and note another two cases in a previous survey. One patient died immediately after the injection of contrast medium but from intraperitoneal haemorrhage. A notable feature of their series is the large number of patients who had cardiac disease.

After this, most authors concentrated on the study of acute reactions, and the allergic responses and delayed reactions were not studied in any detail. Witten et al [2] remarked: "In these, the reaction developed within one hour of injection. No examples of delayed reactions were recognized but no special effort was made to identify reactions of this type".

Davies et al [3] noted four patients whose heart failure worsened after they left the department, which is a delayed reaction due to the fluid overload of the intravascular space, secondary to the administration of a hyperosmolar intravenous contrast medium. Other cases were also observed: ". . . some patients had had delayed rashes after previous urography. In two cases these had recurred after the current urogram, with exactly the same clinical features. Other patients had a 'flu-like' illness on the evening of urography. The occurrence of delayed phlebitis is common knowledge" [4].

Coleman et al [5] noted that iodides cause a 'flu-like' syndrome, parotitis, and skin eruptions, and quoted a case of iodide mumps, following urography, from the *New England Journal of Medicine*, in 1956.

SURVEYS

If it is difficult to be sure of the data when there are multiple observers, it is even more difficult if there is no observer and one has to rely on reports by individual patients. Nevertheless, Panto and Davies [4] conducted a postal survey with an 80% response rate (841 patients) in which they found: "70% of patients had no delayed reactions . . . 13% had arm pain, 5% a rash and 14% had a variety of reactions, many of which were the same as those described in iodism".

McCullough et al [6] studied two large groups of patients given a single injection of either high osmolar (844 patients) or low osmolar (855 patients) contrast medium and again had a response rate of 81%. Acute reactions were more common with the high osmolar contrast medium but delayed rashes, and perhaps parotitis, were apparently more common with the low osmolar medium.

When, however, the data from Panto and Davies [4] were combined with the data of McCullough et al [6] and further unpublished data from the same department, there was no difference between high and low osmolar media in the prevalence of delayed reactions of any type [7]. 1573 patients were given a low osmolar medium with an incidence of delayed rashes of 4.2%, and 2126 a high osmolar agent with an incidence of delayed rashes of 3.4%. This difference is not statistically significant. It arises from the subsets of data because of the small absolute number of reactions, so that samples of some thousands of patients are required to study these rather rare events.

A similar survey on low osmolar contrast media was conducted by Higashi and Katayama [8], who had a 38% response rate (1430 patients); their population included patients having CT as well as urography, but their findings were broadly similar to those of Panto and Davies and McCullough et al; there was no statistical difference between the incidence for CT and urography. They point out that some patients (59) had had similar symptoms within the past year without having had an injection of contrast medium. It is thus uncertain to what extent the contrast medium caused the symptoms. One way of investigating this would be to use other patients attending the Department of Radiology, who had not had an injection of contrast medium, as contemporary controls using a similar postal survey.

Both McCullough et al [6] and Higashi and Katayama [8] reported on the prevalence of parotitis, the latter authors estimating almost the same prevalence as the former, but there is a discrepancy between their text and their tables, the text stating that "One patient had both parotid pain and swelling but he did not receive any therapy and recovered within one day. The other eight patients had painless swelling and all of them cured naturally without medical therapy", while the tables have the numbers and groups reversed, presumably by a transcription error.

Yoshikawa [9] conducted a similar survey on patients having CT and non-ionic contrast medium and achieved an initial 76% response rate, which was increased to 86% by telephoning the non-responders. The results were broadly similar to the findings of other authors, but the author does not comment on parotitis as this was not included in the questionnaire.

COMMENT ON THE SURVEYS

Delayed reactions are common, they occur more often in women and there are a number of well-defined syndromes, which include arm pain, skin rashes, a 'flu-like' illness, variously described gastrointestinal symptoms, pains in various anatomical situations, parotitis and asthmatic attacks.

There does not seem to be any difference in the prevalence of these syndromes in groups of patients given high or low osmolar contrast media.

INDIVIDUAL CASE REPORTS

The following case reports of iodide mumps may be added to those quoted in the papers already cited [10–14]. From the surveys this complication is seen to have a prevalence of about 1% and so can hardly be claimed to be a rare complication; clearly only the more serious cases will end up as case reports. It is usually attributed to free inorganic iodide in the plasma excreted in the saliva, but raised iodide levels are not always associated with parotitis. St Amour et al [15] reported a case of pancreatic mumps with pancreatic swelling (diagnosed by CT) and parotid swelling lasting 4 hours in all. They attributed this to angio-oedema.

A case of fatal vasculitis after urography with iohexol [16], and a fatal Stevens–Johnson syndrome, following the administration of iopamidol in a patient with systemic lupus erythematosis were reported by Savill et al [17].

Toxic epidermal necrolysis which proved fatal after excretory pyelography was reported by Kaftori et al [18] and Lauret et al [19] reported a case of "vegetating iodides", following urography in renal failure in which skin biopsy showed dermo-epidermal necrosis and necrotizing vasculitis. They quote four similar cases found in the literature.

Acute toxic myopathy was reported by Stinchcombe and Davies [20], following an injection of niopam.

There are four case reports of acute thrombocytopenia following intravenous urographic contrast media [21,22]. All recovered, although one recurred on rechallenge. Chang et al [22] point out that it is the duration of thrombocytopenia which is important and "another radiographic examination with the contrast medium may cause bleeding".

Elliott and Reger [23] reported two cases of acute renal failure following low osmolality contrast medium; they claim that this indicates no difference in nephrotoxic potential between low and high osmolar agents. This was the conclusion reached by Dawson in a recent review [24].

THE SPECTRUM OF DELAYED REACTIONS

Most organ systems can be affected by delayed reactions and some have been fatal. A number of the less common reactions can lead to hospitalization and may then, if there appears to be a causal connection, be reported. The experience of the surveys, however, indicates significant underreporting (in, for example, the case of parotitis). A striking feature is the occurrence of delayed reactions in patients with pre-existing disease and in these cases the reactions appear more severe and have sometimes been fatal [15–17].

Pre-existing diseases made worse by contrast media injections include heart failure and renal failure. That acute renal failure can be precipitated in vulnerable individuals by the injection of large doses of contrast media appears to be established.

There is some evidence that asthmatic attacks can be precipitated by injected contrast medium, either acutely [2] or perhaps as a delayed reaction [4,8], but not all asthmatics injected will have such an attack [2].

Skin eruptions of various types have been reported for many years; the severe vasculitic diseases recently reported may merely represent the fatal end of a spectrum which at the other end includes minor rashes and itching.

Gastrointestinal disturbances are common in the general population and those occurring after the injection of contrast medium do not seem to be particularly severe. It seems likely, however, that there is some effect on the gastrointestinal tract.

Delayed arm pain is almost certainly due to thrombophlebitis and is less common with low osmolar media but still, at 10%, a significant reaction. The 'flu-like' illness is slightly more common than this (about 13%) and skin rashes (about 5%) less common; there is about the same incidence of gastrointestinal symptoms (5%), back pain (3%) and headache (3%) are slightly less common, and of the more interesting sequelae, only parotitis (1%) deserves mention, as a relatively frequent delayed reaction.

IODISM

'Iodism' is a relatively poorly defined syndrome, associated with raised plasma inorganic iodide. The severity does not appear to be related to the plasma level and its relevance as a concept to the delayed syndromes occurring after contrast medium injection has been disputed. It is, in any case, possibly better to describe syndromes by neutral names which do not imply an aetiology. Thus, we have 'flu-like' syndrome, the 'serum-sickness-like syndrome' and parotitis ('iodide mumps'). These manifestations all appear to be short-lived and to resolve without sequelae; it appears they are more common in patients with impaired renal function.

McCullough et al [6] comment: "Panto and Davies [4] referred to this syndrome as 'iodism'. This label was attached as it corresponds with dictionary definition. However, this may indicate to some people that the term indicates that the precise causation is known or at least that there is a coherent theory of the aetiology. No such aetiology is implied and the use of the term merely indicates a clinical observation".

Other authors believe there is an aetiology for the skin lesions which is shared with various other iodine compounds [18]. Similarly, parotitis is said to be associated with a raised salivary excretion of inorganic iodine, with perhaps raised serum iodine; this in turn may depend on breakdown of the complex contrast medium molecule to release inorganic iodine [10,14]. However, St Amour et al [15] attributed their case of acute pancreatic and parotid swelling to angio-oedema, and the acute onset (actually during a CT scan) presumably excludes excretion of inorganic iodine as a cause. Davidson et al [11] considered that in their paediatric case with normal renal function the sialadenitis was due to a sensitivity reaction.

Goodfellow et al [16] remark that: "In the case we present there is no direct evidence that the vasculitis was due to iodine sensitivity. It could equally have been due to sensitivity to the entire iohexol molecule".

Thus, there appears to be little consensus on whether these syndromes should properly be grouped together as examples of 'iodism'.

[15] is that there may be different aetiologies for similar syndromes.

CONCLUSIONS

Delayed reactions to urographic contrast medium are more common than acute reactions, often more unpleasant and some have been fatal. There are reports of effects on most organ systems.

The occurrence of delayed reactions cannot be predicted nor can the severity. While there are differences between contrast media, the most important factors undoubtedly reside in the patient. The risk of a fatal outcome in the event of a reaction is increased if the function of any organ system is already compromised by disease, and more so if more than one organ system is compromised.

The detailed study of delayed reactions started about the same time as the introduction of low osmolar contrast media. It should not, therefore, be supposed that the subsequent case reports of delayed reactions associated with the use of these media indicate that patients are more at risk from low osmolar media than when high osmolar media are given (or vice versa).

An apparent increase in delayed rashes with low osmolar media disappeared as more cases were studied, confirming this null hypothesis in a rather unusual and direct way.

REFERENCES

1. Pendergrass HP, Tondreau RL, Pendergrass EP, Ritchie DJ, Hildreth EA, Askovitz SI. Reactions associated with intravenous urography: historical and statistical review. Radiology. 1958; 71: 1–12.
2. Witten DM, Hirsch FD, Hartman GW. Acute reactions to urographic contrast medium. Am J Roentgenol. 1973; 119: 832–840.
3. Davies P, Roberts MB, Roylance J. Acute reactions to urographic contrast media. Br Med J. 1975; 2: 434–437.
4. Panto P, Davies P. Delayed reactions to urographic contrast media. Br J Radiol. 1986; 59: 41–44.
5. Coleman WP, Ochsner SF, Watson BE. Allergic reactions in 10,000 consecutive intravenous urographies. South Med J. 1964; 57: 1401–1404.
6. McCullough M, Davies P, Richardson R. A large trial of intravenous Conray 325 and Niopam 300 to assess immediate and delayed reactions. Br J Radiol. 1989; 62: 260–265.
7. Davies P. Differences between studies of contrast media. (Abstract No. 2173), 17th International Congress of Radiology, Paris, 1989. Abstracts Book, Radiodiagnosis, 344.
8. Higashi TS, Katayama M. The delayed adverse reactions of low osmolar contrast media. Nippon Igaku Hoshasan Gakkai Zasshi. 1990; 50: 1359–1366.
9. Yoshikawa H. Late adverse reactions to non-ionic contrast media. Radiology. 1992; 183: 737–740.
10. Goldberg R, Grosman H, St Louis EL, Gray RR. Contrast induced sialadenitis – a case report. J Otolaryngol. 1984; 13: 331–332.
11. Davidson DC, Ford JA, Fox EG. Iodide sialadenitis in childhood. Arch Dis Child. 1974; 49: 67–68.
12. Koch RL, Byl FM, Firpo JJ. Parotid swelling with facial paralysis: a complication of intravenous urography. Radiology. 1969; 92: 1043–1044.
13. Wylie EJ, Mitchell DB. Iodide mumps following intravenous urography with iopamidol. Clin Radiol. 1991; 43: 135–136.
14. Berman HL, Delaney V. Iodine mumps due to low osmolar contrast media. Am J Roentgenol. 1992; 159: 1099–1100.
15. St Amour TE, McClennan BL, Glazer HS. Pancreatic mumps: A transient reaction to IV contrast media. Am J Roentgenol. 1986; 147: 188–189.
16. Goodfellow T, Holdsock GE, Brunton FJ. Bamforth J. Fatal acute vasculitis after high dose urography with Iohexol. Br J Radiol. 1986; 59: 620–621.
17. Savill JS, Barrie R, Ghosh S, Muhlemann M, Dawson P, Pusey CD. Fatal Stevens–Johnson syndrome following urography with iopamidol in systemic lupus erythematosus. Postgrad Med J. 1988; 64: 392–394.
18. Kaftori JK, Abraham Z, Gilhar A. Toxic epidermal necrolysis after excretory pyelography. Immunologic mediated contrast medium reaction? Int J Dermatol. 1988; 27: 346–347.
19. Lauret P, Godin M, Bravard P. Vegetating iodides after an intravenous pyelogram. Dermatologica. 1985; 171: 463–468.
20. Stinchcombe S, Davies P. Acute toxic myopathy: a delayed adverse effect on intravenous urography with Iopamidol 370. Br J Radiol. 1989; 62: 949–950.
21. Lacey J, Bober-Sarcinelli KE, Farmer LR, Glickman MG. Acute thrombocytopenia induced by parenteral radiographic contrast medium. Am J Roentgenol. 1986; 146: 1298–1299.
22. Chang JC, Lee D, Gross HM. Acute thrombocytopenia after IV administration of a radiographic contrast medium. Am J Roentgenol. 1989; 152: 947–949.
23. Elliott C, Reger M. Acute renal failure following low osmolality radiocontrast dye. Clin Cardiol. 1988; 11: 420–422.
24. Dawson P. On the nephrotoxic potential of the iodinated intravascular contrast agents. Adv X-ray Contrast. 1993; 1: 2–9.

This paper was first published in *Advances in X-Ray Contrast*. 1993; 1: 54–57.

UPDATE

Since the above text was written there has been a small number of new papers on the subject.

The most important is a paper on acute and late adverse reactions to low osmolar contrast media by Mikkonen et al. [1]. The study ran from November 1988 to May 1991 and comprised 4875 consecutive patients from 18 to 19 years of age. The authors were able to show that the incidence of late reactions fell as age increased. Because they had a large number of patients they were able to analyse risk factors for patients having intravenous low osmolar non-ionic contrast. They found that in the acute group the statistically significant risk factors were allergy, drug allergy and previous reaction to contrast medium; the risk factors for late reactions were female sex, allergy, drug allergy and the presence of other diseases, and most importantly previous reaction to contrast medium. Again asthma did not appear to be a risk factor for late reactions.

The same authors [2] went on to study late and acute adverse reactions to iohexol in a paediatric population, the sample size being 321 children under 19 years of age. 50 children had liver disease, heart disease, renal insufficiency or diabetes mellitus. Because of the small size of the sample there was a very small number of patients actually having reactions and the results add little to our knowledge.

Stovsky and his colleagues [3] report what appears to be the first case of a biphasic allergic reaction to a non-ionic agent. The patient had no known allergy, was on no treatment, was not atopic and did not have any environmental allergies. However, 12 hours after the urogram he developed an urticarial reaction and 5 days later developed a moderately severe dermatological eruption which was treated with antihistamine therapy. The authors note that usually there is no correlation between the development of acute and delayed reactions.

Reynolds and colleagues [4] report a case of a 60-year-old woman who developed a severe cutaneous vasculitis 24 hours after the injection of iopamidol and they attribute this to the fact that she was taking hydralazine for renal vascular hypertension at the time. They recommend that urography is contra-indicated if the patient is taking hydralazine. They note that the case is similar to two which were fatal; in one the patient was taking hydralazine and in the other the patient had lupus. It is known that hydralazine can cause a lupus-like syndrome but this was not manifest in the present patient. The authors belive that the simultaneous parotitis supported the suggestion that the iodine compound played a causative role as parotitis commonly accompanies iododerma.

The literature search also found a reference [5] which reported yet again that contrast media can be absorbed across the urothelium in amounts sufficient to cause severe anaphylactoid reactions and the authors report two cases of acute reactions. It must therefore be presumed that such patients are also at risk of delayed reactions.

References

1. Mikkonen R, Kontkanen T, Kivisari L. Acute and late adverse reactions to low-osmolal contrast media. Acta Radiol. 1995; 36: 72–76.
2. Mikkonen R, Kontkanen T, Kivisari L. Late and acute adverse reactions to iohexol in a pediatric population. Pediatr Radiol. 1995; 25: 350–352.
3. Stovsky MD, Seftel AD, Resnick MI. Delayed hypersensitivity reaction after infusion of nonionic intravenous contrast material for an excretory urogram: a case report and review of the literature. J Urol. 1995; 153: 1642–1643.
4. Reynolds NJ, Wallington TB, Burton JL. Hydralazine predisposes to acute cutaneous vasculitis following urography with iopamidol. Br J Dermatol. 1993; 129: 82–85.
5. Weese DL, Greenberg HM, Zimmern PE. Contrast medium reactions during voiding cystourethrography or retrograde pyelography. Urology. 1993; 41: 81–84.

P. Dawson and W. Clauss, (eds.), Advances in X-Ray Contrast: Collected Papers. 34–45
© *1998 Kluwer Academic Publishers.*

Contrast media use in paediatrics

Paul S Babyn, MDCM, and Catherine M Owens, FRCR
The Hospital for Sick Children and the University of Toronto, 555 University Avenue, Toronto, Ontario, Canada

INTRODUCTION

A physician is a person who pours drugs of which he knows little into a body of which he knows less.
Voltaire

The last two decades have witnessed dramatic advances in imaging children. With the advent of the newer imaging modalities, particularly sonography, the applications and need for contrast agents have been altered [1–6]. Techniques such as bronchography and arthrography have nearly been eliminated. Digital fluoroscopy and other technological innovations including improved contrast media safety have reduced the risk of diagnostic investigations, allowing increased application of therapeutic radiological manoeuvres including contrast reduction of intussusception. In this review, we will discuss the current uses of contrast agents in children, their selection, potential adverse reactions and therapy.

CHOICE OF CONTRAST MEDIUM

The contrast agents used in examination of children do not differ from those commonly used in adults, with the choice of contrast medium (CM) determined by its radiological efficacy, side-effects and cost. Extensive experience in children now exists with both conventional, high osmolar contrast media (HOCM), which are ionic, tri-iodinated, fully substituted benzene derivatives, and low osmolar contrast media (LOCM) which are larger molecular dimers with significantly less osmolar composition per gram of iodine [7]. Both types of contrast media have similar radiographic efficacy and are effective for intravascular and other radiological studies [1–7].

Numerous studies have now documented reduction in overall rate of adverse reactions with use of LOCM in childhood [8–12]. LOCM offers a significant advantage in imaging the paediatric patient with its reduction in side-effects, especially nausea and vomiting, as any motion or interruption caused by a minor reaction during examination can be deleterious. Clearly, if costs were equal, LOCM would be preferred. With a nearly five-fold price differential (as currently available in Canada), guidelines restricting LOCM use to those with previous CM reactions, history of severe allergies or asthma, poorly compliant congestive heart failure, increased risk of aspiration, sickle cell disease and very young age have been suggested and are not unreasonable [13]. At our institution, however, we prefer LOCM almost exclusively.

ADVERSE EFFECTS OF CONTRAST MEDIA

The frequency of CM reactions in childhood is similar to that seen in adults for both acute and delayed reactions, with approximately 3–5% of children experiencing an acute reaction [8]. Fortunately, severe reactions are rare. In a survey of paediatric radiologists performing 12,419 intravenous urograms, only five severe reactions were reported [14]. The rarity of adverse reactions is supported by data from the Adverse Drug Monitoring Division of Canada, with only 51 adverse reactions having been reported over 28 years in children aged 0–18 years from intravascular injection of contrast agents (Health Prevention Board Drug Adverse Reaction Program – Report of Suspected Adverse Reactions, 1993). The eleven severe reactions included laryngospasm, profound hypotension, arrhythmias, cyanosis and one death. No matter what the exact incidence of contrast reactions is, both HOCM and LOCM are generally quite safe for intravascular use. The radiologist, however, must not become complacent about the use of CM in children, as adverse reactions including severe reactions will occur notwithstanding use of LOCM.

Two basic mechanisms of toxicity from iodine-containing contrast media exist: chemotoxicity and anaphylactoid or allergic-type reactions [15]. Chemotoxic reactions are due to specific physicochemical effects of the CM and, unlike anaphylactoid reactions, are directly dependent upon its dose, concentration, osmolality, potential calcium ion binding and nature of the associated cation [15]. The rate and site of injection (e.g. intravenous vs intraarterial) therefore play a significant role in chemotoxicity [7]. Chemotoxic reactions, particularly those due to the osmolality of the injected CM, usually affect the central nervous system, respiratory tract, cardiovascular system and kidneys and, in general, are similar to those seen in adults [1,2].

Careful attention is necessary to *individualize* the dose of contrast agent to suit the child because of the wide physical variations encountered from premature infant to young child to adolescent. Accidental overdose of contrast agent has been reported, especially in infants undergoing urography with HOCM [16,17]. The administered amount was often quite substantial, approaching the lethal dose found in animal work. Following overdosage, apnoea, cyanosis, brain and pulmonary oedema and pulmonary haemorrhage may rapidly develop [16–18]. Treatment should consist of vigorous supportive therapy and possibly dialysis to remove the remaining circulating CM. Pulmonary oedema may also occur without significant overdosage having been given [19–20]. Generally, a strict adherence to a dosage based upon body weight is suggested for intravascular administration.

Although the exact incidence and mechanism of nephrotoxicity are disputed, renal dysfunction can occur in children following CM administration. It is more likely to be encountered in those with diabetes, pre-existing renal disease and following excessive dosage of CM [21–24]. It may also occur in other situations, such as recently reported in a set of xiphaomphalapagus conjoined twins where 2 cc/kg (presumed weight of one twin) of LOCM were given [25].

The osmolar effects of CM, particularly HOCM, can be considerable, especially in young, debilitated

neonates (Table 1). Injection of 4 cc/kg of HOCM will raise serum osmolality by 5% and circulating blood volume by 20% in a young infant [26]. Rapid fluid shifts into the vascular space compensate for the hypertonic CM leading to hypertonic dehydration of the body tissue [26]. Compensatory changes in organs such as the central nervous system occur with resultant increased CSF and cerebral venous pressure [3,27]. Seizures and neurotoxicity may occur following myelography or intravenous administration of CM and are more likely in neonates with pre-existing neurological disorders [3]. Rapid fluid shifts out of the vascular system due to osmolar effects may follow ingestion of large amounts of hypertonic CM, which will be discussed later.

Table 1. Osmolality of commonly used intravenous contrast agents in the paediatric population (modified from Ref 2)

Contrast agent	Trade name	Iodine (mg I/ml)	Osmolality (mOsm/kg)	Relative to serum osm.
Diatrizoate	Hypaque 60	280	1500	~5×
	Hypaque 18	85	450	~1.5×
Iothalamate	Conray-60	280	1500	~5×
Iohexol	Omnipaque 300	300	700	~2×
Iopamidol	Isovue 300	300	600	~2×
Ioxaglate	Hexabrix	320	600	~2×

Small subcutaneous extravasations of CM may occur during intravenous administration of contrast agents, with a frequency of occurrence ranging from 0.01–0.17%. They are usually trivial and need no specific treatment. If large extravasations occur, however, they may cause severe local complications with skin and subcutaneous tissue necrosis [28,29]. Skin debridement, grafting and plastic reconstruction may rarely be needed [28].

In experimental animal work supported by human experience, extravasation of conventional ionic media produces more significant reactions than LOCM [29]. Young infants are at particular risk of extravasation, especially when pre-existing cannulas are used or when cannulas are obscured by overlying tape and bandages. Pressure (power) injectors should be used with caution in sedated or young children. Use of LOCM is suggested when potential for extravasation exists [29].

Anaphylactoid-type reactions

Anaphylactoid-type reactions are probably less important than chemotoxicity (especially osmolality) as a major cause of morbidity or mortality in childhood [1,2]. In patients with a history of CM reactions, multiple allergies or asthma, an increased risk of reaction is present. In these children, pretreatment with a combination of corticosteroids and/or antihistamines and use of LOCM are recommended if an alternative diagnostic evaluation is not possible [15]. Caution and careful consideration are warranted in those children predisposed to tumour lysis following steroid administration (e.g. acute leukaemia and lymphoma), as renal failure may occur [30].

TREATMENT OF ADVERSE REACTIONS

Despite all precautions, adverse reactions still occur. Anticipation and careful planning before a reaction occurs are crucial to the immediate management of severe reactions and patient survival [15]. Frequent monitoring of vital signs, ensuring a patent airway and establishment of intravenous access must be promptly instituted with administration of specific appropriate therapy (Table 2).

CURRENT APPLICATIONS OF CONTRAST AGENTS

GENITOURINARY

Intravenous pyelography

Although intravenous urography is now much less frequently needed, it is still valuable in assessment of renal calculi and where anatomical delineation of the renal pelvis and ureters is required. In children, good excretory urograms can be obtained with a bolus intravascular injection of iodinated contrast material of 1 ml/kg (up to a maximum of 50 ml).

Over the last few years, there has been a general acceptance of a lower contrast agent dosage than previously utilized. Our suggested guidelines are shown in Table 3.

Table 3. Recommended contrast agent dosage for intravenous urography

Weight (kg)	Dosage (ml)
0–5	2–3
6–15	15
16–20	20
21–25	25
26–35	30
36–45	35
46–55	40
56–65	45
66+	50

(Assumes normal renal function and use of iohexol 300 or equivalent.)

Voiding cystourethrography (VCUG)

Voiding cystourethrography remains an important method of evaluating the lower urinary tract in children, especially for those with suspected structural or functional abnormalities including vesicoureteric reflux. The use of radionuclide cystography should not be overlooked, especially for follow-up studies, as the radiation dose is substantially reduced.

For cystography, the traditional HOCM agents, diatrizoate (Hypaque or Renografin) or iothalamate (Cysto-conray) are generally used. Prepackaged bottles generally contain between 18 and 30% iodine. A considerably lower concentration, including an isotonic 10% iodine, can be utilized with a much lower anticipated rate of complications [31]. There is no substantive evidence at this time that the lower osmolar agents are less toxic to the bladder than HOCM at cystography.

The quantity of CM administered varies according to the patient's age, information desired and status of the urinary tract. A sufficient volume must be given to ensure prompt micturition, but overdistension should be avoided.

Cystitis or systemic complications due to absorption of the CM from the bladder can occur, albeit uncommonly.

Table 2: Therapy for acute contrast reactions

Type of reaction	Signs and symptoms	Therapy	Dosage and route of administration	Comments	
Minor	**Nausea/vomiting**				
	Transient	Supportive			
	Severe, prolonged	Prochlorperazine injectable	>2 years old: 0.1–0.15 mg/kg intramuscular or intravenous	Administer slowly IV – caution drowsiness	Can be repeated every 6 hours
	Urticaria				
	Scattered, transient	Supportive			
	Scattered, prolonged	Diphenhydramine injectable (Benadryl)	1.25 mg/kg IV, intramuscular	Observe – for drowsiness	Can be repeated every 2–3 hours
	Profound	Ranitidine injectable	1.25–1.9 mg/kg IV in 24 h	Administer slowly and reduce dosage with renal impairment	Administer in divided doses every 6 hours
Moderate	**Bronchospasm**				
	Mild–moderate	Oxygen	3 L/min via mask		
		Subcutaneous epinephrine 1:1000	0.01 mg/kg to max of 0.2 mg subcutaneously	Dose Limit Premature neonates (0.05 mg) Full-term neonates 1 ml (0.1 mg)	Can be repeated every 15 minutes. Maximum 2 doses
	Wheezing, prolonged	Metaproterenol (Alupent)	1–2 inhalations via metered dose	Ensure adequate inhalation	Every 4–6 h
Severe	**Bronchospasm**				
	Accelerating, severe	Intravenous epinephrine 1:10,000	0.01 mg/kg to 0.1 mg maximum IV	Administer slowly	Every 15 minutes. Maximum 2 doses
	Hypotension and sinus rhythm or tachycardia	Intravenous fluids	10-20 ml/kg IV (rapid bolus)	Caution fluid overload Monitor blood pressure and urine input	5% albumin, normal saline Ringer's lactate
	Hypotension and bradycardia	Intravenous fluids +	10-20 ml/kg IV (rapid bolus)		
		Atropine	0.02 mg/kg to 0.60 mg (max) IV	May be given via endotracheal tube if IV route not possible	Every 20 minutes, as required
	Seizures/convulsions				
	Isolated	Intravenous fluids			
	Multiple or continuous	Diazepam	0.1–0.3 mg/kg IV	Dose Limit <5 y: 5 mg/dose >5 y: 10 mg/dose	Every 20 minutes, as required × 3 dosages
				May cause hypotension and apnoea when given IV. Administer slowly	

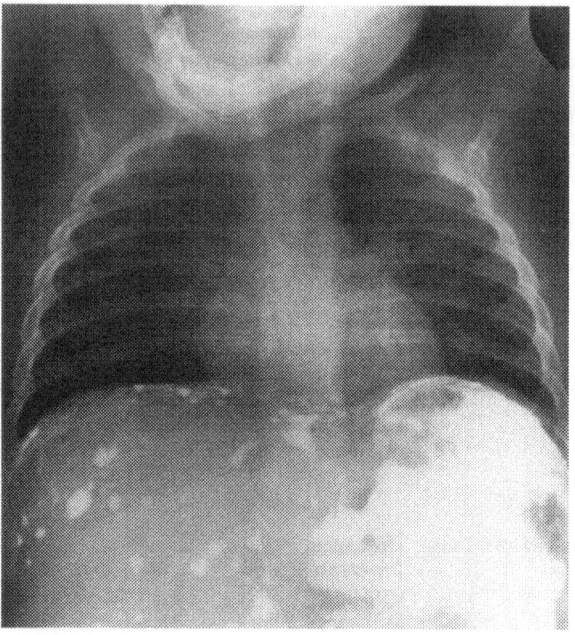

(a)

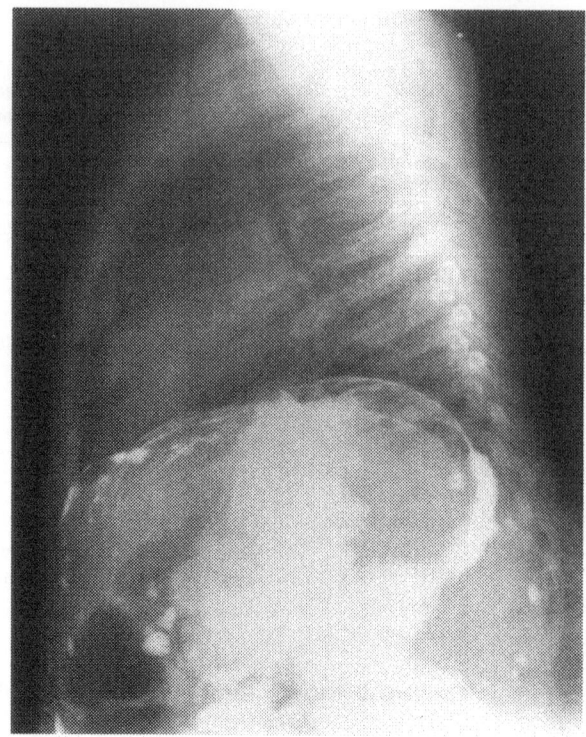

(b)

Figure 1 Frontal (a) and lateral (b) chest radiographs demonstrating extensive peritoneal barium extravasation from previous attempted intussusception reduction

GASTROINTESTINAL TRACT

Choice of contrast agent

The choice of contrast agent for GI studies in children must be tailored to the clinical issue at hand. Single contrast studies of the GI tract are all that are required in neonates and in infancy where congenital anomalies or gross pathology are most likely. Double contrast studies of the alimentary tract are reserved for those children suspected of having specific mucosal abnormalities such as inflammatory bowel disease.

Air is a very safe and effective CM for evaluation of many congenital GI abnormalities. For example, when necessary, insufflation of a small amount of air, usually 5–10 ml, is sufficient to diagnose oesophageal atresia and differentiate this from retropharyngeal perforation. Barium is the standard contrast agent for imaging the GI tract, and has a long history of effective use. Barium is cheap, non-absorbable, inert and insoluble, and since it is hypotonic when administered, does not cause significant osmolar fluid shifts. In addition, barium is an excellent X-ray beam attenuator with high contrast. Although there are few studies directly comparing barium and other agents in children, one such study reports far better visualization, in vitro, of 50% weight per volume barium [32].

Occasionally there are significant disadvantages in using barium, some directly related to barium itself with others due to the preservatives used in its preparation. 'Barium' is actually a suspension of barium particles and stabilizing agents needed to ensure its suspension. The barium particles will eventually settle and flocculate, so that studies requiring visualization of bowel for more than about 6 hours cannot be well performed using barium.

Barium may cause impaction in patients with a partial colonic obstruction [33,34], or (rarely) of the small bowel in patients with cystic fibrosis.

In part due to its additives, barium is toxic in the peritoneum. Over the years these additives have changed and barium has become less toxic (Figure 1).

The duration of exposure to barium is also

important. Acutely, barium causes an inflammatory exudate which may cover the particles so that within a few hours it may not be possible to wash the peritoneal cavity clear [32]. Exudates of fluid into the peritoneum may cause problems with hypovolaemia in patients who are already unstable [35]. Eventually, the inflammation [33] caused by the barium will result in formation of granulomas and adhesions [32,36]. A similar inflammatory reaction may be seen in the pleural space or mediastinum, with granulomata and fibrotic adhesions developing [33,37].

In young infants in whom there is a probability of barium retention within the colon, water may be absorbed, causing fluid overload [38,39].

When aspirated, small amounts of barium are usually easily tolerated. Aspiration of larger amounts, however, may prove fatal [33,36]. It is probable that the volume of aspirated material is more important than the nature of the material in these situations. One infant died 3 months after aspiration of barium, with acute and chronic inflammation and barium aggregation observed at autopsy [38]. While the resultant inflammatory granulomas are better tolerated in the lung than in the abdomen, they are best avoided.

In most infants and young children, the choice of barium preparation is not very crucial. Suggested ranges of barium concentration for varying studies, however, are as follows: double-contrast upper GI studies, 200–250% weight per volume; single-contrast upper GI studies, 50–100% weight per volume; small bowel follow-through studies, 50–100% weight per volume; enteroclysis studies, 20–30% weight per volume; double-contrast colon studies, 15–20% weight per volume. Double-contrast or enteroclysis studies rarely need to be performed in young infants but are more frequently required in older children [4].

WATER-SOLUBLE CONTRAST MATERIAL

Water-soluble contrast materials are of value in the evaluation of the GI tract in specific instances, particularly where perforation or fistula formation is suspected. It is important, especially in neonates, to use near isotonic solutions of CM for GI studies, as rapid fluid shifts from the body tissues and vascular system into the bowel lumen occur.

Any of the HOCM can be sufficiently diluted to obtain an isotonic solution. Unfortunately, this results in a very low iodine concentration, in the range of 40–80 mg iodine per ml. Though images with these dilute solutions may be diagnostic (especially in small infants when enemas are administered with a low kilovoltage of about 50 kV), the image quality is usually very poor [40]. LOCM can be utilized in an isotonic solution and still contain a high enough concentration of iodine to provide excellent images of the bowel in young children [41,42]. When used in such solutions, the oral administration of these agents has no effect on the patient's haematocrit level or serum osmolality [43]. Better visualization of the bowel can be achieved with LOCM [44], as less dilution of the contrast occurs. In addition, unlike HOCM, LOCM do not damage the bowel mucosa [45]. HOCM are potentially toxic to bowel mucosa, presumably because of their osmolality. The deaths of three infants from necrotizing enterocolitis following the use of Gastrografin or Renografin-76 enemas to treat meconium ileus have been reported [46,47]. As HOCM are partially absorbed from the bowel, their presence in the blood vessels will cause further shifts of fluid from the extravascular space into the vessels, increasing blood volume [48] and further dehydrating body cells. When aspirated, HOCM are extremely toxic and can cause pulmonary oedema [32,33,36] or lung inflammation [49] with possible fatal consequences.

In the one large study of the use of LOCM in 115 infants and young children, aspiration occurred on 13 occasions, seven of which were categorized as massive. In each case, the contrast agent was cleared rapidly from the lungs and all the patients did well [6].

Although most of the early work with LOCM has been done with metrizamide, more recent work has been done with iohexol which confirms the advantages of LOCM.

LOCM can result in excellent bowel visualization many days after administration of the agent [42], and are less irritating to body tissues such as the peritoneal cavity or the lung than barium and hypertonic water-soluble agents [44,49]. Seven infants who had leakage of these agents into the peritoneal cavity suffered no side-effects [6].

We recommend LOCM in those situations where a strong probability of aspiration of contrast material into the lungs exists, including: (i) swallowing disorders,

40

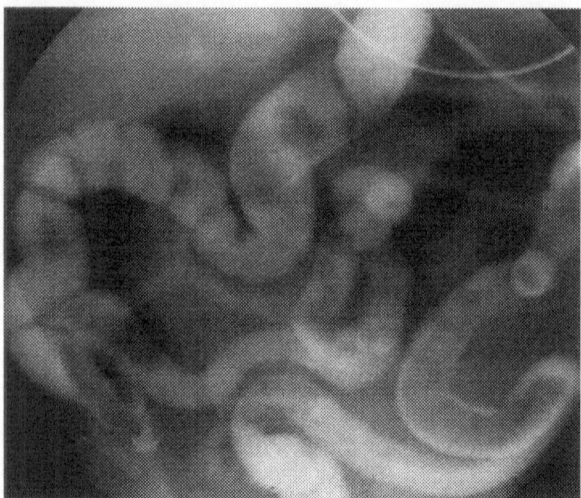

Figure 2 Supine anteroposterior abdominal radiograph demonstrating a small meconium-filled colon with reflux of Hypaque 18 into meconium-filled small bowel, in a patient with meconium ileus and cystic fibrosis

especially in patients with CNS impairment; (ii) probable oesophageal obstruction; (iii) tracheooesophageal fistula; and (iv) selected instances of vomiting in infants and children where the likelihood of aspiration is great. In the last situation it must be emphasized that this should represent only a small percentage of children with vomiting, as, generally, barium remains the contrast agent of choice.

In addition, in those clinical situations where a significant risk of CM leakage from the bowel is likely, LOCM should be used. Such situations include necrotizing enterocolitis, possible bowel perforation, unexplained pneumoperitoneum and the gasless abdomen where severe bowel disease is likely. Metrizamide has been successfully used to identify bowel perforation in a number of infants [50], with several having had fairly localized bowel perforations, including some without evidence of pneumoperitoneum. When using LOCM, bowel perforation can be recognized either by identification of the contrast agent within the peritoneal cavity or by its appearance within the genitourinary tract [50]. In neonates presenting with abdominal distension and delayed or absent passage of meconium, LOCM are also of value.

In general, the major drawback and consideration in use of LOCM is their cost, especially if large volumes are needed. In our opinion, especially in the neonate where perforation is a possibility, LOCM should be employed, as most studies can be accomplished with as few as 20 ml of contrast material, the cost of which is insignificant relative to the possible complications.

THERAPEUTIC USES OF CONTRAST IN THE GI TRACT

Meconium ileus

In 1969, Noblet [51] reported non-operative relief of uncomplicated meconium ileus by Gastrografin enema. The success of this method has been widely confirmed by other authors and it has been generally accepted as a gratifying alternative to surgical intervention which may carry a high post-operative mortality risk [52]. Complications of Gastrografin enemas are mainly related to the high osmolality of this substance, which causes depletion of plasma water, leading to hypovolaemia and hypertonicity. These side-effects have led to the replacement of Gastrografin by less toxic agents such as Hypaque 18 (Figure 2).

It has been suggested that the high osmolality of Gastrografin was the important element in therapy of meconium ileus, and in the past, undiluted CM was recommended. However, when advanced into massively dilated proximal loops, Gastrografin is likely to remain stagnant. Deaths have been reported from prolonged exposure to the irritant effects on the bowel mucosa, with consequent disruption of the mucosal barrier, leading to enterocolitis and sepsis. Localized induced haemoconcentration of small bowel capillaries by hypertonicity may result in sluggish blood flow and bowel ischaemia [53]. The systemic effects of hypertonicity due to rapid fluid shifts from plasma to interstinal lumen and transient osmotic diarrhoea can result in depletion of plasma water, leading to hyperalbuminaemia, and cardiovascular compromise.

Intussusception reduction

Intussusception is a common paediatric condition that has for many years been diagnosed and treated with the use of fluoroscopically guided barium enemas. Liquid

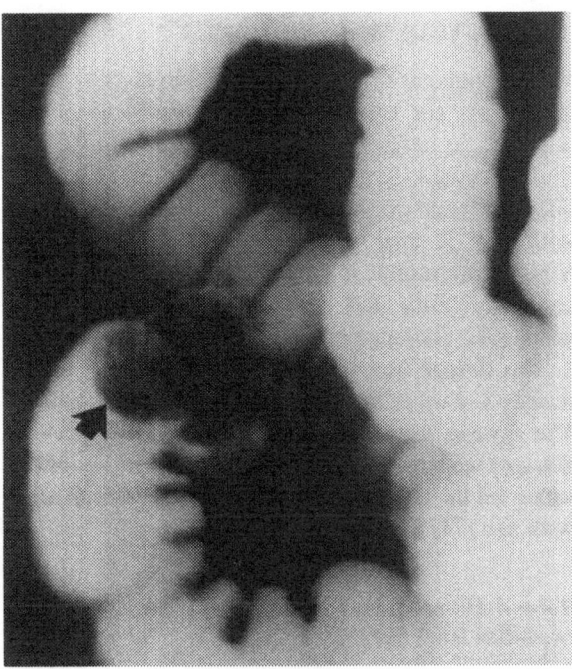

Figure 3 AP radiograph during barium reduction, outlining intussusceptum (arrow)

preparations, particularly those with barium, have been the most frequently used contrast agents in North America [54,55] (Figure 3).

Air has long been used in China and its use in North America and Europe is increasing. Following appropriate hydration and initial stabilization, reduction rates with barium range from 53–85% with most authors reporting rates of 70–85% [56,57]. Similar reduction rates have been reported with air, ranging from 57–96% [55,56], with many authors reporting rates greater than 90%. Comparative, non-randomized studies have shown reduction rates to be 1–32% higher with air than with barium [58]. Claims of greater efficacy, lower radiation exposure and increased safety when perforation does occur [58–60] have resulted in the recent increase in the use of air enemas in the United States and Canada.

Although radiological reduction is now accepted as the most desirable method of initial treatment, there is no consensus regarding its applicability to patients with intussusception and signs of small bowel obstruction. Several authors consider long-standing symptoms and small bowel obstruction as contraindications to hydrostatic reduction attempts [61–63], pointing out the danger of bowel perforation and peritonitis [64,65]. Many authors state that intussusceptions associated with small bowel obstruction often prove reducible and that the attempt is relatively harmless. We recommend an initial trial of reduction for all patients, with the exception of those in shock or with signs of peritonitis, who should proceed directly to surgery [66–68]. We recognize the potential for perforation, especially in those less than 3 months of age and with symptoms persisting over 2 days. Close liaison with clinicians is vital and the radiologist's role in intussusception reduction is probably valid in these cases, as long as there is immediate surgical back-up.

ANGIOGRAPHY AND ANGIOCARDIOGRAPHY

Contrast administration for angiography and angiocardiography must adhere carefully to the osmolar limitation guidelines. As multiple injections of CM are frequent, studies must be carefully planned so that a dose limitation of 3 ml/kg body weight is adhered to while simultaneously obtaining sufficient diagnostic information. LOCM cause significantly fewer haemodynamic alterations in systemic systolic blood pressure, dP/dT, heart rate, ventricular and diastolic pressures, and prolongation of the QT interval.

Young children are particularly vulnerable to changes in cardiac rhythm and pump function as a consequence of the chemotoxic effects of contrast agents. Immediate adverse reactions include dysrrhythmias, alterations in heart rate and blood pressure, as well as increased vascular resistance and decreased myocardial contractility [10].

For good visualization of the arterial, capillary and venous circulation, the total volume of CM must be given rapidly. The concentration of the CM, however, can be substantially reduced in digital angiography compared with conventional film technique. Thus, isotonic concentrations of CM can be used in digital studies.

BODY COMPUTED TOMOGRAPHY

Intravenous contrast enhancement for computed tomography (CT) in children is an invaluable adjunct in many situations. In our institution, a bolus injection of CM is followed by rapid sequence scanning, rather than an intravascular drip. In general, a dosage of 2 ml/kg body weight, up to a maximum of 120 ml, is adequate for most paediatric patients. We use LOCM because of their reduced local irritation and pain on bolus injection, as well as the lower incidence of nausea and vomiting. The higher CM cost is a trade-off for the improved patient throughput and lack of patient motion, resulting in better-quality scans. With the lower body fat content of children, especially infants, we find that a satisfactory bowel opacification can be obtained by using a 1% weight/volume solution of barium or iodinated CM. The contrast should be administered approximately 1–2 hours prior to the scan, allowing bowel transit to opacify the colon, and an additional oral bolus is administered just prior to the scan in order to opacify the stomach and proximal small bowel. A contrast enema may be required in selected cases, utilizing the same CM dilution.

CENTRAL NERVOUS SYSTEM

CT

In neuroradiological practice, the use of iodinated contrast agents for CT relies on the breakdown of the blood–brain barrier. Our regular paediatric dose for CT has been 3 ml/kg body weight of 300 mg of iodine per ml up to a maximum of 100 ml, while other authors prefer to use 2 ml/kg. Again, as described previously, the CM is given in a bolus. It has been shown that the use of an increased dose in the CNS gives preferential enhancement of lesions with little increase in normal tissues [3]. CM are used routinely in evaluation of seizure disorders and, on rare occasions, to localize the tentorium and the falx for anatomical purposes. CM should be used with caution in the neonate with pre-existing neurological disorders, as seizures may occur [3].

(COMPUTED) MYELOGRAPHY

With the advent of MR imaging there are now fewer indications for myelography and ventriculography; however, in cases of isolated tethered cord (or tight filum terminale syndrome) and in patients who have difficulty in co-operating, myelography in conjunction with CT is still a useful diagnostic tool [69]. Ventriculography is still used in analysing the communication and compartmentalization of the ventricular cisterns [70].

The dosage used for ventriculography is approximately 1–2 ml of isotonic non-ionic contrast agent. The dosage used for myelography, unlike previous dosages we have discussed, is not based on weight but rather on the capacity of the CSF space, which varies with age [71] (Table 4).

Table 4. **Dosage table for complete myelography (modified from Ref 71)**

Age	Concentration (mg I/ml)	Volume (ml)
Less than 2 months	180–210	1/2
2 m–2 y	180–210	2–4
3 y–7 y	180–210	4–8
8 y–12 y	180–210	7–12
13 y–17 y	180–210	8–14

Ionic CM should not be used in myelography because of their high neurotoxicity. Even the first non-ionic contrast agent, metrizamide, is more neurotoxic than the newer contrast agents, such as iohexol, iopamidol or ioversol [72–75]. Despite different reports by various authors, no significant differences have been shown to exist between newer non-ionic contrast agents [75,76].

The incidence of adverse drug reactions following intrathecal usage of non-ionic contrast agents varies from 26–44% [72,77,78], with the commonest reactions being headaches and vomiting. It is often uncertain whether or not the reaction is caused by the dural puncture itself or the contrast agent used. No seizures have been recorded in our experience with the newer non-ionic contrast agents, but they have been reported by others [27,79], with an incidence of approximately 1 per 300 with metrizamide.

BRONCHOGRAPHY

Bronchography is now seldom utilized in North America or Western Europe in infants and children, as alternative diagnostic methods, in particular CT and bronchoscopy, are now available and often superior. In addition, there is now increased recognition of the complications of bronchography, especially in neonates and young children.

Most bronchograms in infants and children are performed with the patient under general anaesthesia because of the inability of the child to co-operate. This creates some technical problems, in that the usual spontaneous respiratory efforts to distribute bronchographic contrast are substituted by the positive pressure application of the anaesthetist. Segmental or lobar collapse is common (as can be seen with general anaesthesia alone) and fatalities have been noted [80]. All agents used for bronchography have two problems, the first a direct toxic effect on the bronchi and lung tissue, and the second direct mechanical effects on the airway. Mechanical airway obstruction is a particular problem in infancy because of the small size of the infants' airways.

Oil-based agents such as dionosil have a long history of use in paediatric bronchography. They have a very high viscosity and it is therefore possible that small or even moderate-sized fistulas would not be visualized with these agents. The low osmolar agents have been used for bronchography in infants and children with mixed results. Metrizamide in a concentration of 220 mg I/ ml has been clinically used without evident harm [80]. Nevertheless, there is experimental animal evidence that even such agents do damage tracheal epithelium [81].

In the infrequent instances where paediatric bronchography is still required, low osmolar agents are perhaps preferred over dionosil because of lack of mechanical airway obstruction.

CONCLUSION

Recent developments have all centred on the non-ionic contrast agents because they are less toxic than the ionics; however, the cost of the non-ionics is significantly higher than that of the ionics. Despite the cost debates, in our institution we routinely use non-ionic contrast agents for all intravascular and intrathecal studies in the paediatric age group. We do so because of the lower complication rate of minor side-effects including extravasation, whilst admitting there is no convincing evidence of any decrease in the mortality rate.

It is vital to remember, with the proliferation and sophistication of imaging modalities now available and the continued rapid progress in imaging, that close consultation between radiologist and clinician is of increasing importance, to minimize cost and patient radiation exposure and maximize diagnostic benefit. It is this team approach which will improve the care given to the sick children in our practices and best make use of improvements in contrast agents.

GENERAL REFERENCES

1. Franken Jr EA, Sato Y, Smith WL. Pediatric use of contrast agents. In: Skucas J, ed. Radiographic Contrast Agents. Gaithersburg, Maryland: Aspen Publishers, 1989: 462–485.
2. Smith WL, Franken EA. Pediatric contrast agents. In: Katzberg RW, ed. The Contrast Media Manual. Baltimore, Maryland: Williams and Wilkins, 1992: 207–221.
3. Chuang S. Contrast agents in pediatric neuroimaging. AJNR. 1992; 13: 785–794.
4. Cohen MD. Choosing contrast media for the evaluation of the gastrointestinal tract of neonates and infants. Radiology. 1987; 162: 447–456.
5. Zerin JM. Contrast studies of the gastrointestinal tract in the neonate. Semin Pediatr Surg. 1992; 1: 284–295.
6. Ratcliffe JF. The use of low osmolality water soluble (LOWS) contrast media in the pediatric gastro-intestinal tract: a report of 115 examinations. Pediatr Radiol. 1986; 16: 47–52.

SPECIFIC REFERENCES

7. Moires TW, Katzberg RW. Intravascular contrast media: properties and general effects. In: Katzberg RW, ed. The Contrast Media Manual. Baltimore, Maryland: Williams and Wilkins, 1992: 1–18.
8. Hitoshi Katayama MD, et al. Adverse reactions to ionic and non-ionic contrast media – a report from the Japanese committee on safety of contrast media. Radiology. 1990; 175: 621–628.
9. Di Sessa TG, Zednikova M, Hinaishi S, Jarnakani JM, Higgins CB, Friedman WF. The cardiovascular effects of metrizamide in infants. Radiology. 1983; 148: 687–691.

44

10. Pelech AN, Allarel SM, Hard RT, Giddins NG, Collins GF. A comparison of iohexol and diatrizoate-meglumine in children undergoing cardiac catheterization. Invest Radiol. 1991; 26: 665–670.

11. Carro JJ, Trindade E, McGregor M. The risks of death and of severe nonfatal reactions with high- vs. low-osmolality contrast media: a meta-analysis. AJR. 1991; 156: 825–832.

12. Lawrence V, Matthai W, Hartmaier S. Comparative safety of high-osmolality and low-osmolality radiographic contrast agents. Invest Radiol. 1992; 27: 2–28.

13. Levin DC, Gardiner GA, Karasick S, et al. Cost containment in the use of low osmolar contrast agents: effect of guidelines, monitoring and feedback mechanisms. Radiology. 1993; 189: 753–757.

14. Gooding CA, Berdon WE, Brodeur AE, Rowen M. Adverse reactions to intravenous pyelography in children. AJR. 1975; 123: 802–804.

15. Bush W, Swanson DP. Acute reactions to intravascular contrast media: types, risk factors, recognition and specific treatment. AJR. 1991; 157: 1153–1161.

16. Ansell G. Fatal overdose of contrast medium in infants. Br J Radiol. 1970; 43: 395–396.

17. Kassner EG, Elquezabal A, Pochaczebsky R. Death during intravenous urography. Overdosage syndrome in young infants. New York State J Med. 1973; 73: 1958–1966.

18. Junck L, Marshall W. Fatal brain edema after contrast-agent overdose. AJNR. 1986; 7: 522–525.

19. Berdon WE. Pulmonary edema in infants who receive contrast materials for urograms (letter). Radiology. 1981; 139: 508.

20. Wood BP, Smith WL. Pulmonary edema in infants who receive contrast materials for urograms (letter). Radiology. 1981; 139: 508.

21. Barrett BJ, Carlisle EJ. Meta-analysis of the relative nephrotoxicity of high- and low-osmolality iodinated contrast media. Radiology. 1993; 188: 171–178.

22. Kathali RE, Taylor GJ, Woods WT, et al. Nephrotoxicity of non-ionic low-osmolality vs. ionic high-osmolality contrast media: a prospective double-blind randomized comparison in human beings. Radiology. 1993; 186: 183–187.

23. Dawson P. On the nephrotoxic potential of the iodinated intravascular contrast agents. Adv X-Ray Contrast. 1993; 1: 2–9.

24. Berg KJ, Jakobsen JA. Nephrotoxicity related to X-ray contrast media. Adv X-Ray Contrast. 1993; 1: 10–18.

25. Boo NY, Mahmud MN, Samad SA. Radiocontrast-induced nephropathy in a pair of xiphaomphalapagus conjoined twins during the neonatal period. Acta Pediatr Scand. 1991; 80: 735–737.

26. Cohen MD. Intravenous urography in neonates and infants. What dose of contrast should be used? Br J Radiol. 1979; 52: 942–944.

27. Maly P, Back-Gansmo T, Elmqvist D. Risk of seizures after myelography: comparison of iohexol and metrizimade. AJNR. 1988; 9: 879–883.

28. Elam E, Dorr R, Lagel K, Pond G. Cutaneous ulceration due to contrast extravasation: experimental assessment of injury and potential antidotes. Invest Radiol. 1991; 26: 13–21.

29. McAlister WH, Kissane JM. Comparison of soft tissue effects of conventional ionic, low osmolar ionic and non-ionic iodine containing contrast material in experimental animals. Pediatr Radiol. 1990; 20: 170–174.

30. Luna-Fineman S, Healy MV, Parker BR. Corticosteroid pretreatment for potential contrast reactions in children with lymphoreticular cancer: a word of caution. AJR. 1990; 155: 357–358.

31. Grossman H, Merten D, Effman E, Plucinsky R. Isotonic water soluble contrast material for cysto-urethrogram. J Urol. 1982; 128: 1006–1008.

32. Foley MJ, Ghahremani GG, Rogers LF. Reappraisal of contrast media used to detect upper gastrointestinal perforations: comparison of ionic water-soluble media with barium sulfate. Radiology. 1982; 144: 231–237.

33. Gelfan DW, Ott DJ. Gastrointestinal contrast agents. In: Taveras JM, Ferrucci JT, eds. Radiology. Vol. 4. Philadelphia: Lippincott, 1986: 1–7.

34. Margulis AR, Burhenne HJ, eds. Alimentary tract roentgenology. 3rd ed. St. Louis: Mosby, 1983: 87–108.

35. Eklof O, Wald J, Thomasson B. Barium peritonitis: Experience of five pediatric cases. Pediatr Radiol. 1983; 13: 5–9.

36. Dodds WJ, Stewart ET, Vlyment WJ. Appropriate contrast media for evaluation of oesophageal disruption. Radiology. 1982; 144: 439–441.

37. Ginai AZ, Kate FJW, Berg RGM, Hoornstra K. Experimental evaluation of various available contrast agents for use in the upper GI tract in case of suspected leakage; Effects on mediastinum. Br J Radiol. 1985; 58: 585–592.

38. McAlister WM, Siegel MJ. Fatal aspiration in infancy during gastrointestinal series. Pediatr Radiol. 1984; 14: 81–83.

39. Kirks DR. Practical pediatric imaging: diagnostic radiology of infants and children. Boston: Little, Brown, 1984: 536, 650–652, 665–666.

40. Kuhns LR, Kanellitsas C. Use of isotonic water-soluble contrast agents for gastrointestinal examinations in infants. Radiology. 1982; 144: 411.

41. Coussement A. Non-ionic and dimeric contrast agents (letter to editor). Radiology. 1983; 148: 318–319.

42. Ratcliffe JF. The use of ioxaglate in the paediatric gatrointestinal tract: a report of 25 cases. Clin Radiol. 1983; 34: 579–583.

43. Clarke E, Siefle RL. Effect of oral metrizamide on hematocrit and serum osmolality in the neonate. Invest Radiol. 1984; 19: 599–600.

44. Ginai AZ. Experimental evaluation of various available contrast agents for use in the gastrointestinal tract in case of suspected leakage: Effects on peritoneum. Br J Radiol. 1985; 58: 969–978.

45. Schwartzenruber DJ, Bicumire DF, Cohen M, Block Gunter M, Grosfeld JL. Use of iohexol in the radiographic diagnosis of ischemic bowel. J Pediatr Surg. 1986; 21: 525–529.

46. Leonidas JC, Burry VF, Fellows RA, Beatty EC. Possible adverse effect of methylglucamine diatrizoate compounds on the bowel of newborn infants with meconium ileus. Radiology. 1976; 121: 696–698.

47. Grantmyre EB, Butler GJ, Gillis DA. Necrotizing enterocolitis after Renografin-76 treatment of meconium ileus. AJR. 1981; 136: 990–991.

48. Dawson P, Grainger RG, Pitfield J. The new low-osmolar contrast media: a simple guide. Clin Radiol. 1982; 34: 221–226.

49. Ginai AZ, Kate FJW, Berg RGM, Hoornstra K. Experimental evaluation of various available contrast agents for use in the upper gastrointestinal tract in case of suspected leakage: Effects on lungs. Br J Radiol. 1984; 57: 895–901.

50. Cohen MD, Weber TR, Grosfeld JL. Bowel perforation in the newborn: diagnosis with metrizamide. Radiology. 1984; 150: 65–69.

51. Noblet HR. Treatment of uncomplicated meconium ileus by Gastrografin enema: A preliminary report. J Pediatr Surg. 1969; 4: 190–197.

52. Donnison AB, Shwachman H, Gross RE. A review of 164 children with meconium ileus seen at the Children's Hospital Medical Center, Boston. Pediatrics. 1966; 37: 833–850.

53. Lutzger LG, Factor SM. Effects of some water-soluble contrast media on the colonic mucosa. Radiology. 1976; 118: 545–548.

54. Campbell JB. Contrast media in intussusception. Pediatr Radiol. 1989; 19: 293–296.

55. Meyer JS. The current radiology management of intussusception: A survey and review. Pediatr Radiol. 1992; 22: 323–325.

56. Sweh YH, Ting WH, Yeh HH. Reduction of intestinal intussusception in infancy by colonic air insufflation. Chin Mgo J. 1964; 83: 666–673.

57. Eklof OA, Johanson L, Lohr G. Intussusception: Hydrostatic reduction and incidence of leading points in differentiate. Pediatr Radiol. 1980; 10: 83–86.

58. Palder SB, Ein SH, Stringer DA, Alton D. Intussusception: Barium or air? J Pediatr Surg. 1991; 26: 271–275.

59. Gu L, Alton DJ, Daneman A, et al. Intussusception reduced in children by rectal insufflation of air. AJR. 1988; 150: 1345–1348.

60. Leonidas JC. Treatment of intussusception with small bowel obstruction: Application of decision analysis. AJR. 1985; 145: 665–669.

61. Ein SH, Stephens CA. Intussusception: 354 cases in 10 years. Pediatr Surg. 1971; 6: 16–27.

62. Auldist AW. Intussusception in a children's hospital: a review of 230 cases in seven years. Aust NZ J Surg. 1970; 40: 136–148.

63. Franken EA, Smith WL, Chernish SM, Fletcher BD, Goldman H. The use of glucagon in hydrostatic reduction of intussuception – a double blind study of 30 patients. Radiology. 1983; 107: 590–601.

64. Humphrey A, Ein SH, Mok PM. Perforation of the intussuscepted colon. AJR. 1981; 137: 1135–1138.

65. Ein SH, Mercer S, Humphrey A, Macdonald P. Colon perforation during attempted barium enema reduction of intussusception. Pediatr Surg. 1981; 16: 313–315.

66. Bjamason G, Peterson G. The treatment of intussusception: thirty years experience of Gothenburg's Children's Hospital. Pediatr Surg. 1968; 3: 19–23.

67. Frye TR, Howard WHR. The handling of ileocolic intussusception in a pediatric medical center. Radiology. 1970; 97: 187–191.

68. Williams HJ. Intussusception facts, fallacies and practicals. Minn Med. 1975; 58: 140–147.

69. Chuang S, Hochhauser L, Harwood-Nash D, Armstrong D, Burrows P, Savoie J. The tethered cord syndrome. Paper presented at XIII Symposium Neuroradiologicalum, Stockholm, Sweden, 1986.

70. Chuang S, Fitz CR, Harwood-Nash D. The use of metrizamide ventriculography in pediatric hydrocephalus. Paper presented at the Society for Pediatric Radiology, San Francisco, CA, USA. 1981.

71. Harwood-Nash DC, Fitz CR. Myelography: neuroradiology in infants and children. Vol 3. St. Louis: Mosby, 1976: 1125–1166.

72. Bannon KR, Braun IF, Pinto R. Comparison of radiographic quality and adverse reactions in myelography: iopamidol and metrizamide. AJNR. 1983; 4: 312–313.

73. Moschini L, Manara O, Bonaldi G. Iopamidol and metrizamide in cervical myelography, side-effects, EEG, and CSF changes. AJNR. 1983; 4: 848–850.

74. Trevisan C, Malaguti C, Manfredini M. Iopamidol vs. metrizamide myelography: clinical comparison of side-effects. AJNR. 1983; 4: 306–308.

75. Witwer G, Cacayorin ED, Bernstein AD. Iopamidol and metrizamide for myelography: prospective double-blind clinical trial. AJR. 1984; 143: 869–873.

76. Davies AM, Evans N, Chandy J. Outpatient lumbar radiculography; comparison of iopamidol and iohexol and a literature review. Br J Radiol. 1989; 62: 716–723.

77. Meyer JS, Daneman BC, Buonamo C, Bemin JA. Air and liquid contrast agents in the management of intussusception: A controlled randomized trial. Radiology. 1993; 188: 507–511.

78. Miller DL, Chang R, Wells WT, et al. Intravascular contrast media: effect of dose on renal function. Radiology. 1988; 167: 607–611.

79. Lipman JC, Wang A, Brooks ML, et al. Seizure after intrathecal administration of iopamidol. AJNR. 1988; 9: 787–788.

80. Smith W, Franken EA. Metrizamide as a contrast agent for visualization of the tracheobronchial tree: Its drawbacks and possible advantages. Pediatr Radiol. 1984; 14: 158–160.

81. Wells WD, Burbmage MD. Direct effects of contrast media on rat lungs. Can Assoc Radiol J. 1991; 42.

This paper was first published in *Advances in X-Ray Contrast*. 1993;1:74–85.

P. Dawson and W. Clauss, (eds.), Advances in X-Ray Contrast: Collected Papers. 46–51
© 1998 Kluwer Academic Publishers.

Cardiac use and effects of contrast agents

Michael A Bettmann, MD
Dartmouth Hitchcock Medical Center, Department of Diagnostic Radiology, Lebanon, New Hampshire, USA

A consideration of the cardiac effects of contrast agents is important for several reasons. First, all contrast agents have some measurable effect on the heart and cardiovascular system when injected, regardless of site of injection. Secondly, injection of contrast agents at high concentrations and high doses both in and around the heart is common for examinations including the now rare intravenous DSA studies, pulmonary angiography and coronary artery interventions. Thirdly, the most severe adverse reactions encountered with contrast agents all have cardiac components, and many such reactions may, in fact, be secondary to direct cardiac effects. Finally, a substantial body of work exists examining the effects of contrast agents on various aspects of cardiac function. Such studies shed substantial light on the adverse effects of contrast agents in general.

EXPERIMENTAL EFFECTS

The effects of contrast agents on the heart have been examined in many different experimental models over several decades. The major problems in defining the cardiac effects of contrast agents are that, firstly, multiple diverse models are available for evaluation, and secondly, multiple factors are operative in the effects of contrast agents on the heart. For example, the effect of contrast infusion into the left ventricle differs among different species, and effects within a species are different if a classical isolated heart preparation is utilized, compared with either an open-chest model or a closed-chest model. There are also multiple mechanims by which the contrast agent can have a specific effect. For example, effects on left ventricular function must be considered not only in terms of direct effects when the contrast agent is injected into the left ventricle, but also in terms of secondary effects, related to perfusion of the contrast agent through the coronary arteries, or of reflex effects related to action of

the contrast agents on the peripheral arteries. It is also very difficult to separate the effects on the conduction system from those on contractility. Perhaps most importantly, there are multiple factors operative in the effects of contrast agents. In addition to hypertonicity, types of ion present, if any, play a role, as do additives; the duration of exposure, which is in part a function of viscosity, is another consideration.

Given this multiplicity of operative factors, and the difficulty of relating experimental results to clinical findings, it is not surprising that the picture that emerges is not entirely clear. One finding which is beyond dispute is that the single most important factor overall is osmolality. The higher the osmolality to which the heart is exposed, the more marked the effects. As shown in one study, however, the effects of hypertonicity depend on the site exposed [1]. Utilizing an isolated rat heart model, it was shown that a non-ionic hypertonic control solution had a mild positive inotropic effect, but caused coronary artery vasodilatation, leading to a decrease in perfusion pressure; an ionic hypertonic control solution, on the other hand, had a negative inotropic effect, while still causing similar dilatation of the coronary arteries. When contrast agents were examined in the same model, all contrast agents, both ionic and non-ionic, caused a basically osmolality-dependent negative inotropic effect, as well as a decrease in perfusion pressure. This suggests that factors other than osmolality are involved. A non-ionic, dimeric formulation, iotrolan, despite its isotonicity, produced a mild positive inotropic effect, with a slight decrease in coronary artery resistance. What these findings demonstrate is that osmolality alone has a positive effect on myocardial contractility directly and leads to coronary artery dilatation. The addition of ions produces no additional effect on the coronary arteries in this model, but converts the direct positive inotropic effect to a negative one. When contrast agents are used clinically, they depress both myocardial contractility and coronary artery tone, in direct relationship to the degree of hypertonicity, regardless of whether or not ions are present.

Not surprisingly, the results are somewhat different when other models are utilized. This is partly because different parameters are measured, different types of contrast agents are studied and different control substances are employed [2,3].

The effects of contrast agents on the conduction system have also been examined in many different models, in an attempt to isolate these effects. The results are again confusing. Some studies suggest that low osmolality contrast agents which are totally lacking in ions lead to a greater likelihood of significant adverse effects on the conduction system [4–7]. Some studies also suggest that the addition of sodium, or preferably sodium and calcium ions, decreases the risk of such adverse effects on the conduction system compared with non-ionic agents alone. This issue is complicated, not only by differences from model to model, but also by a widespread belief that the use of low osmolality agents is inherently safer than the use of high osmolality agents, and the addition of ions not only

changes the formulation to an ionic one, but increases the osmolality to a certain extent. It is perhaps most important to keep in mind that experiments are designed to achieve a better understanding of the actions of contrast agents, and any findings from them may not be directly applicable to the clinical situation.

Relevant to cardiotoxic effects are the observations that contrast agents chelate calcium, that calcium plays a major role in both conduction and contractility, and that a sudden decrease in contractility, as occurs clinically on intracoronary contrast agent injection, may lead to electromechanical dissociation [8]. This may lead to a clinical picture which mimics an acute anaphylactoid response. However, although the aetiology is different, the initial treatment, aimed at restoring cardiac function, is identical.

The specific effects of calcium binding have been examined experimentally in several studies. In one model, it was evident that calcium binding occurred, and that the magnitude of this was less with low than with high osmolality agents and was, at least in part, related to the additives in contrast agent formulations [9] (Figures 1–3). This is not surprising, as both EDTA and sodium citrate are known to sequester calcium avidly.

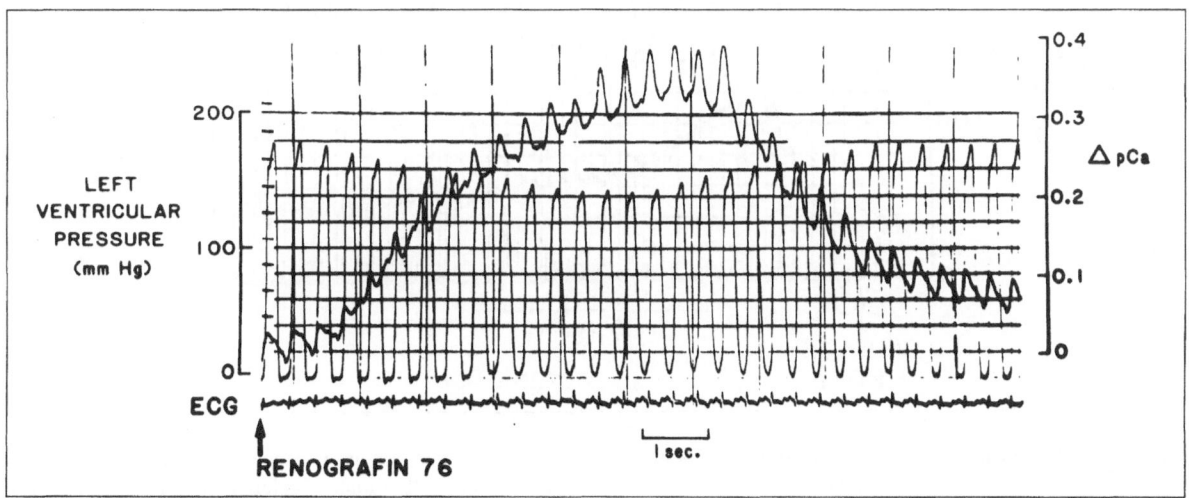

Figure 1 Selective canine left coronary artery injection of sodium meglumine diatrizoate (Renografin), 370 mg I/ml, demonstrates a fairly marked fall in peak systolic pressure and a marked alteration in ECG morphology beginning within 3 s of injection. This is accompanied by a discrete rise of the left ventricular end diastolic pressure as well as a marked increase in pCa, which indicates a fall in ionized calcium.

48

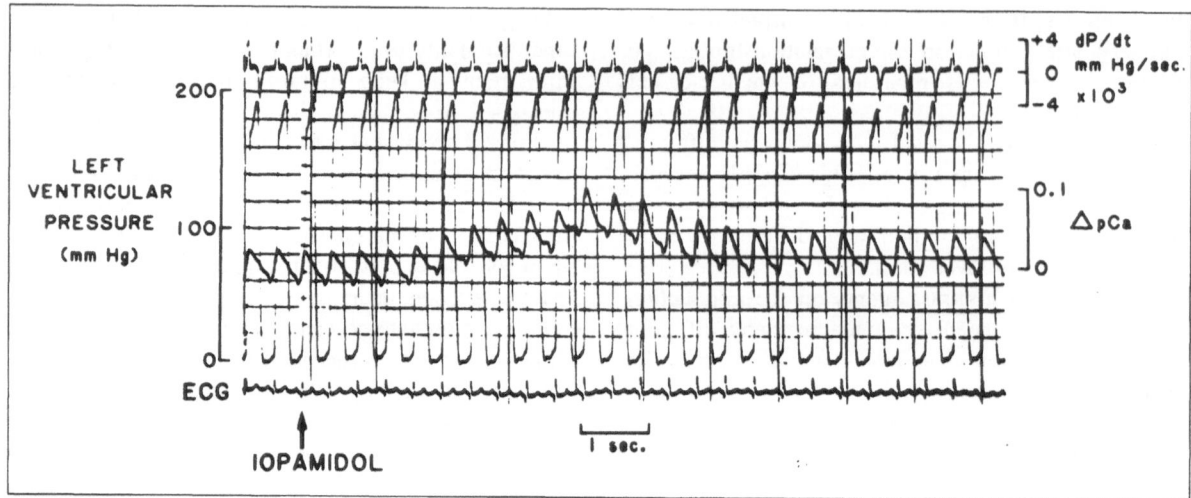

Figure 2 Selective injection of iopamidol, 370 mg I/ml, demonstrates changes which are similar in character to those observed with a high osmolality agent, but the changes are of markedly decreased magnitude. Note a minor fall in peak systolic pressure, a minor change in ECG morphology and a minor fall in ionized calcium. Although minor, this latter alteration may still have clinically important sequellae.

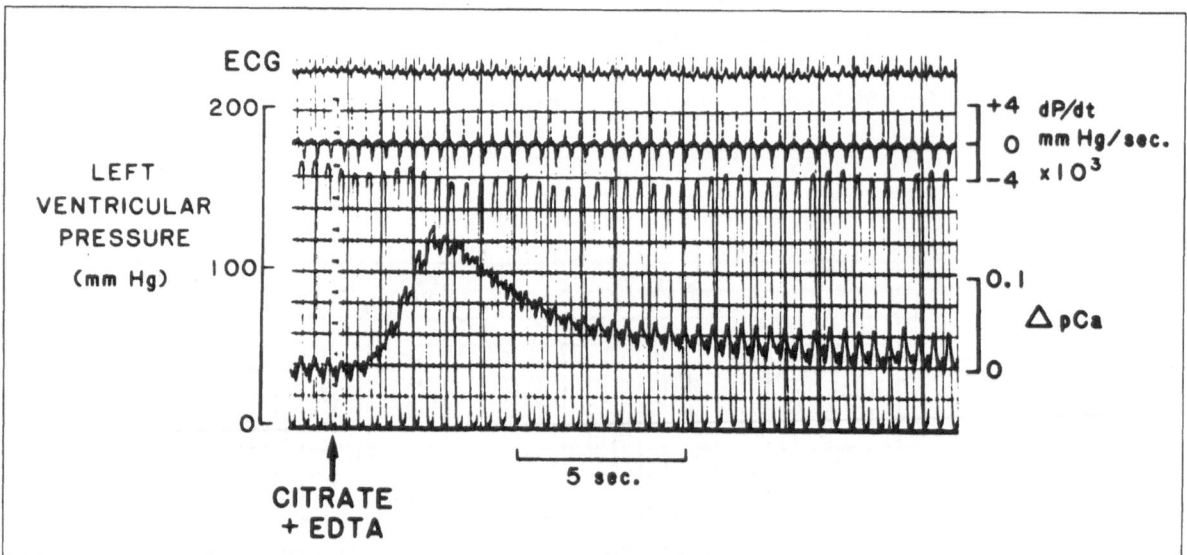

Figure 3 Selective injection of sodium citrate and sodium EDTA in normal saline, in concentrations identical to those which exist in Renografin 76. There is a minor fall in peak systolic pressure and little change in the electrocardiogram, but a marked fall in ionized calcium. This again, may have substantial clinical relevance.

strate that there are substantial effects of contrast agents on cardiac function. Furthermore, there are several key factors which play a role in these effects, including hypertonicity and the presence of certain ions, and several key effects, namely, suppression of myocardial contractility, dilatation of coronary arteries, and modification of the pattern of conduction. These effects are complex and clearly deserve further investigation. Perhaps most importantly, experimental findings in various models must be interpreted in the light of the vast clinical experience with contrast agents.

CLINICAL EFFECTS

Numerous large and small studies have examined the clinical effects of contrast agents. Most of these have compared high and low osmolality agents. There is universal consensus, based on these studies and on experimental studies, that low osmolality contrast agents have fewer and less marked effects on the heart. The difficulty emerges, however, in defining the magnitude and the importance of such differences. Overall, there has really been no evidence to suggest that the quality of the images achieved is dependent on anything other than the iodine concentration injected. In general, an iodine concentration ranging from 320 to 370 mg I/ml is both necessary and sufficient for cardiac imaging.

The haemodynamic effects of contrast agents are fairly clear: on injection into the left ventricle, there is a decrease in left ventricular peak systolic pressure with a delayed increase in heart rate, presumably related to reflex effects. On selective coronary injection, there is a fall in both pressure and rate which is almost immediate (3–5 s) and returns to normal within no more than 60 s. These effects are clearly greater and more prolonged with high osmolality contrast agents. Some studies have attempted to differentiate between different low osmolality contrast agents, with somewhat conflicting results [10,11].

Perhaps the key consideration is the relative incidence of serious adverse effects with different contrast agents, varying in formulation or in type of molecule. One study examined high-risk patients [12] and found that, although there was little difference in minor side-effects, there was a significant difference in favour of the low osmolality agent in regard to major side-effects. Although the study is now relatively old and was a small one, it provides support for belief in the increased safety of low osmolality agents in high-risk cardiac patients, during cardiac studies. Another, larger study had similar findings, with a lower incidence of both contrast-related and overall adverse events which were considered severe, with a non-ionic agent compared with a high osmolality agent. Interestingly, however, as has emerged also from many other studies, there was no difference in mortality rate between the two contrast agents [13].

Other clinical studies have concentrated on specific factors, such as arrhythmias [14] and thrombosis [15–17]. Regarding ventricular fibrillation, which experimental studies suggested might occur more frequently with non-ionic agents, compared with either ionic agents [5] or even non-ionic agents with ions added [6], clinical experience suggests that the incidence of this life-threatening complication, if it changes at all, decreases with the use of low osmolality agents in general [14].

Thrombosis related to contrast use is a complex issue. Experimental studies clearly suggest that there is a difference between ionic (high and low osmolality) and non-ionic agents, with a more marked retardation of coagulation with ionic agents. There are real questions as to the clinical relevance of these findings. Various studies have attempted to clarify the question of modification of clotting by different types of contrast agent. This is clearly most important in a setting such as coronary angiography or, even more so, coronary angioplasty, where the physician is dealing with a catheter in a small vessel which is likely to have both relatively slow flow and endothelial abnormalities, both of which predispose to thrombosis. The most reasonable conclusion that can be arrived at is that there may be differences between ionic and non-ionic agents with respect to retardation of coagulation but, in a clinical setting, the importance of such differences appears to be outweighed by other factors. That is, the specific contrast agent is likely to be less important as an isolated risk factor than other risk factors, such as the underlying condition of the artery, flow, and angiographic technique. Differential effects of contrast agents on thrombosis are also unlikely to have any particular relevance on the venous side.

Overall, it is clear that low osmolality contrast agents are very safe from the cardiac point of view. The exact incidence of clinical adverse reactions is difficult to determine due to the multiplicity of variables, including specific contrast agent type, technique and underlying patient risk factors. The major concern with cardiac use of contrast agents clearly relates to the direct cardiac effects, although other adverse events, such as cutaneous or pulmonary manifestations or renal failure, may also occur. Many patients undergoing coronary angiography have other vascular disease, hypertension or diabetes mellitus, with the result that underlying renal dysfunction is likely to be present in a substantial proportion. Contrast-related renal failure, therefore, is often a consideration. Nonetheless, even among patients with a history of prior contrast reaction, the major concern is a direct cardiac one. Experimental studies indicate that the single most important factor in regard to the cardiac effects of contrast agents is the underlying cardiac status; the more compromised the overall cardiac function, the greater the likelihood of a significant adverse reaction. In the setting of significant cardiac disease, there is a clear advantage in using low osmolality contrast agents.

COST AND UTILIZATION CONSIDERATIONS

Several authors have addressed the question of overall safety, with particular reference to serious cardiac events and mortality [18–20]. Individual studies clearly demonstrate differences in haemodynamic responses to high and low osmolality contrast agents. Larger studies, however, whether formal meta-analyses, prospective randomized studies or reviews of multiple studies, appear to suggest that, in cardiac use, while there is a clear difference in the incidence of mild and moderate reactions, and while there is probably a difference (of uncertain magnitude) in the incidence of severe reactions, there is no difference in the mortality related to the contrast agent [21,22].

Several large, prospective studies have found that there is no evidence of financial advantage with universal use of low osmolality agents, despite a lower incidence of adverse events [23,24]. These, and other studies, suggest that the presence of specific cardiac risk factors, notably unstable angina, increases the likelihood of an adverse reaction, and can be used as

an indicator for cost-effective selective use of low osmolality contrast agents. The cost differential between high and low osmolality contrast agents in any case differs in different countries. At least in the United States, and probably elsewhere, because of universally increasing concern about rising healthcare costs, a strategy of selective use of low osmolality contrast agents has been advocated. Strategies for such use have been specifically outlined by the American College of Cardiology [25] for cardiac catheterization. A dispassionate philosophical explanation of such a position has been well enunciated by individuals involved in technology assessment and healthcare reform [26]. Although some individuals have strong opinions, which are reflected not only in individual standards of practice but also in national policies in several countries, many questions remain to be answered, specifically with regard to the aetiology and the significance of cardiac effects of contrast agents.

CONCLUSIONS

With every intravascular administration of a contrast agent, cardiac effects occur. These effects are clearly most marked with injection into or near the heart. These cardiac effects vary widely in their clinical manifestations, from insignificant but measurable haemodynamic alterations to ventricular fibrillation and cardiac arrest. In attempting to define the cardiac effects of contrast agents, endpoints are relatively easy to measure, in the form of haemodynamic or conduction alterations. The importance of these effects is, however, difficult to define. The most important, single factor in cardiac contrast agent-related adverse events is osmolality. In the clinical setting, the underlying state of the patient, and specifically of her cardiac function, appears to be far more important than the contrast agent itself. Substantial work remains to be done in defining the aetiology of adverse cardiac effects. Such work is difficult because of the multiplicity of factors involved, because of differences between various experimental models, and because of unresolved questions concerning the relevance of such experimental models to the clinical situation. Osmolality does play a major role in clinical adverse events, but calcium binding is also clearly but calcium binding is also clearly important. The relevance of the presence or absence of small amounts of added ions remains uncertain at this time.

In terms of clinical utility, many physicians advocate universal use of low osmolality contrast agents for cardiac applications. In broadest terms, when attempting to think in terms of overall benefit to society, the most rational approach currently is selective use of low osmolality agents in patients who are at increased risk of adverse cardiac events, such as those with unstable angina, severe or acute CHF, pulmonary hypertension, severe valvular disease or significant conduction abnormality. Whilst much remains uncertain, the introduction of low osmolality ionic and non-ionic contrast agents certainly represents a major advance.

REFERENCES

1. Fleetwood G, Bettmann MA, Gordon JL. The effects of radiographic contrast media on myocardial contractility and coronary resistance; osmolality, ionic concentration, and viscosity. Invest Radiol. 1990; 25: 254–260.
2. Higgins CB, Sovak M, Kelley MJ, Newell JD. Direct myocardial effects of intracoronary administration of new contrast materials with low osmolality. Invest Radiol. 1980; 15: 39–46.
3. Klow NE. Low osmolality contrast media and metabolic changes in the myocardium during free and reduced coronary flow in the dog. Invest Radiol. 1990; 25: 127–132.
4. Kozeny GA, Murdock DK, Euler DE, et al. In vivo effects of acute changes in osmolality and sodium concentration on myocardial contractility. Am Heart J. 1985; 109: 290–296.
5. Morris TW. Ventricular fibrillation during right coronary arteriography with ioxaglate, iohexol and iopamidol in dogs. Invest Radiol. 1988; 23: 205–208.
6. Donadieu AM, Cardinal HA, Bonnemain B. Incidence of ventricular fibrillation during coronary arteriography in the rabbit: a comparative study of isotonic ioxaglate and iohexol. Invest Radiol. 1987; 22: 106–110.
7. Baath L, Besjahov J, Ohsendal A. Sodium–calcium balance in non-ionic contrast media. Invest Radiol. 1993; 28: 223–227.
8. Caulfield JB, Zir L, Harthorn JW. Blood calcium levels in the presence of arteriographic contrast material. Circulation. 1975; 52: 119–123.
9. Bourdillon PD, Bettmann MA, McCraeken S, et al. Effects of a new non-ionic and a conventional ionic contrast agent on coronary sinus ionized calcium and left ventricular hemodynamics in dogs. J Am Coll Cardiol. 1985; 6; 845–853.
10. Scholtz ME, Svatos VJ, Myburgh DP. Comparison between a low-osmolar ionic (ioxaglate) and a low-osmolar non-ionic (iopamidol) contrast agent in cardiac imaging. S Afr Med J. 1988; 73: 168–171.
11. Hirshfeld JW, Jr, Wieland J, Davis CA, et al. Hemodynamic and electrocardiographic effects of ioversol during cardiac angiography: comparison with iopamidol and diatrizoate. Invest Radiol. 1989; 24: 138–144.
12. Feldman RL, Jalowiec DA, Hill JA, Lambert CR. Contrast media-related complications during cardiac catheterization using hexabrix or renografin in high-risk patients. Am J Cardiol. 1988; 61: 1334–1337.
13. Hill JA, Winniford M, Cohen MB, et al. Multicenter trial of ionic versus non-ionic contrast media for cardiac angiography. Am J Cardiol. 1993; 72: 770–775.
14. Missri J, Jeresaty RM. Ventricular fibrillation during coronary angiography: reduced incidence with non-ionic contrast media. Cathet Cardiovasc Diagn. 1990; 19: 4–7.
15. Vacek JL, Hibiya K, Rosamund TL, et al. Anticoagulant effect of iohexol vs. ioxaglate during cardiac catheterization. J Invas Cardiol. 1992; 4: 139–144.
16. Gasperetti CM, Feldman MD, Burwell CR, et al. Influence of contrast media on thrombus formation during coronary angioplasty. J Am Coll Cardiol. 1991; 18: 443–450.
17. Hwang MH, Piaoz E, Murdock DK, et al. Risk of thromboembolism during diagnostic and interventional cardiac procedures with non-ionic contrast media. Radiology. 1990; 174: 453–457.
18. Hirshfield JW Jr. Low-osmolality contrast agents – who needs them? N Engl J Med. 1992; 326: 482–484.
19. Caro JJ, Trindade E, McGregor M. The risks of death and of severe nonfatal reactions with high- vs. low-osmolality contrast media: a meta-analysis. AJR. 1991; 156: 825–832.
20. Lawrence V, Matthai W, Hartmaier S. Comparative safety of high-osmolality and low-osmolality radiographic contrast agents. Invest Radiol. 1992; 27: 2–28.
21. Steinberg EP, Moore RD, Gopalou R, et al. Safety and cost effectiveness of high-osmolality as compared with low-osmolality contrast material in patients undergoing cardiac angiography. N Engl J Med. 1992; 326: 425–430.
22. Barrett BJ, Parfrey PS, Vavasour HM, et al. A comparison of non-ionic, low-osmolality radiocontrast agents with ionic, high-osmolality agents during cardiac catheterization. N Engl J Med. 1992; 326: 431–436.
23. Hlatky MA, Morris KG, Pieper KS, et al. Randomized comparison of the cost and effectiveness of iopamidol and diatrizoate as contrast agents for cardiac angiography. J Am Coll Cardiol. 1990; 16: 871–877.
24. Powe NR, Moore RD, Steinberg EP. Adverse reactions to contrast media: factors that determine the cost of treatment. AJR. 1993; 161: 1089–1095.
25. American College of Cardiology Cardiovascular Imaging Committee. Use of non-ionic or low osmolar contrast agents in cardiovascular procedures. J Am Coll Cardiol. 1993; 21: 269–273.
26. Eddy DM. Broadening the responsibilities of practitioners; the team approach. JAMA. 1993; 269: 1849–1855.

This paper was first published in *Advances in X-Ray Contrast*. 1994;2:2–7.

P. Dawson and W. Clauss, (eds.), Advances in X-Ray Contrast: Collected Papers. 52–56
© 1998 Kluwer Academic Publishers.

Contrast agents in interventional radiology

Peter Dawson, PhD, MRCP, FRCR
Royal Postgraduate Medical School, Hammersmith Hospital, London, UK

INTRODUCTION

There has been a proliferation and refinement in recent years of a variety of imaging modalities and techniques such as computerized tomography (CT), CT angiography (CTA), magnetic resonance imaging and spectroscopy (MRI/MRS), magnetic resonance angiography (MRA), ultrasound imaging (US), and duplex Doppler ultrasound (DD). These frequently allow non-invasive diagnostic procedures to be performed, but in interventional radiology, by definition, an invasive technique has to be used and, more often than not, a contrast-enhancing agent is necessary. This is so whether the radiologist is working in the cardiovascular system, the biliary tree or the renal tract, to take three common examples. It is entirely fair to say that without the availability of effective, well-tolerated, and basically safe, water-soluble X-ray contrast agents, interventional radiology would have barely developed, let alone attained the important central role it currently occupies in the management of many patients.

In the last decade, the introduction of non-ionic contrast agents has been a boon to radiologists in general, and perhaps to interventional radiologists in particular [1,2]. Furthermore, as will be discussed, the development of digital systems for angiography has been associated with a number of advantages, not least of which is the requirement for smaller contrast medium concentrations and total doses [3]. These complementary developments of non-ionic contrast agents and of digital systems are of the utmost importance since many complex procedures entail the use of numerous contrast medium injections and potentially very high total body loads of these (mild) tissue poisons. Furthermore, many patients undergoing contrast-assisted interventional procedures may be ill with multisystem disease or organ failure [2] and, though modern agents are benign and well tolerated by any usual standards of drug assessment, any toxicity

assumes a much greater importance in such patients, who may also, further to complicate the situation, be on regimens of a multiplicity of other drugs. Poor cardiac reserve, cardiovascular instability and impaired renal function are common problems in these patients [2]. Those with arteriovenous malformations may be hypervolaemic and in high-output cardiac failure or have raised pulmonary arterial pressure.

Some interventional procedures, in the hands of some operators, will also involve more exotic treatment of contrast agents, such as mixing them in vitro (or effectively in vivo) with other drugs or agents such as dextrose, ethanol, antibiotics or lignocaine, or may involve subjecting them to extremes of physical conditions for which they were not designed, such as boiling prior to use or exposing them to the high energy densities associated with the use of lasers.

CONTRAST AGENTS – CHEMISTRY AND PHARMACOLOGY

The chemistry and pharmacology of the water-soluble contrast agents has been extensively reviewed and will not be repeated here [1,4,5]. In summary, the conventional high osmolality contrast agents are ionic, monomeric, benzene-ring-based structures; current low osmolality agents are of two types: non-ionic, monomeric agents, represented by iohexol, iopamidol, iopromide and ioversol, and ionic, dimeric agents, represented only by sodium meglumine ioxaglate.

An incoming generation of agents, the non-ionic dimers [6], which are iso-osmolar with plasma, will probably soon be the agents of choice in all high-dose procedures.

Some formulations of the ionic agents contain sodium cations and sodium citrate or sodium EDTA additives. These bind calcium, which is an important factor in their cardiotoxicity [7]. Sodium meglumine ioxaglate contains sodium, but, like the non-ionic contrast agents, has a calcium rather than a sodium EDTA additive, which results in little reduction of circulating calcium levels. The ioxaglate cation is, however, quite an effective calcium binder itself. The non-ionics contain the calcium EDTA preparation and, by definition, no sodium. The non-ionic molecules themselves are very weak calcium binders.

All these compounds have essentially the same pharmacokinetics, being distributed rapidly though the extracellular fluid space of the body and excreted by passive glomerular filtration with no active tubular secretion or reabsorption.

The agents possess toxicity by virtue of hyperosmolality [8], their chemical structure (chemotoxicity) [4] and their additives [7,9]. The hyperosmolality contribution is significantly reduced in the non-ionic monomers and in the ionic dimer and is completely eliminated in the non-ionic dimers [6]. The chemotoxicity is mediated by a combination of Coulomb forces and non-specific hydrophobic interactions between biological macromolecules and the hydrophobic portions of the contrast molecules, principally the benzene ring–iodine system [5]. The low clinical toxicity of the non-ionic contrast agents stems from the combination they offer of a relatively low osmolality, low chemotoxicity, an absence of sodium ions, and a non-calcium-reducing form of EDTA. All contrast agents have been described as mild tissue poisons. To the extent that this is true, the non-ionic agents, and particularly the non-ionic dimers, represent the mildest of the available poisons.

IDIOSYNCRATIC REACTIONS

Idiosyncratic/anaphylactoid reactions are not a problem specific to interventional radiology and the same principles apply as in diagnostic radiology [10]. The reactions may occur, unpredictably, following the administration of any of these agents [11]. Though unpredictable, such reactions are known to be more likely to occur in certain patient groups: in those who have reacted on a previous occasion to a contrast agent, in those with established allergies to other drugs and agents, and in asthmatic and atopic individuals [10]. In these patients, non-ionic contrast agents should be used as a matter of course. Such reactions are essentially

dose-independent, may occur following even a subcutaneous or intradermal test dose and may occasionally occur during non-vascular procedures such as PTC or hysterosalpingography, presumably as a result of intravasation. In patients definably at risk, it is certainly worth considering the use of a non-ionic agent, even for these non-vascular procedures.

The use of corticosteroid prophylaxis in at-risk patients is controversial and, on close examination, the evidence to support its widespread practice is unconvincing [12].

HIGH-DOSE TOXICITY

While much energy has been expended on the problem of idiosyncratic/anaphylactoid reactions to contrast agents, particularly with regard to the elucidation of their underlying mechanisms, the prediction of their likelihood and their prophylaxis and management, much less attention has been paid to the problem of the dose-dependent side-effects of the agents. In terms of frequency, these are of greater importance than idiosyncratic reactions in a number of patient groups who are well represented in interventional radiology, such as those with poor cardiac reserve, impaired renal or hepatic function and multisystem organ failure [2]. Even without such obvious underlying risk factors, many patients may be at risk from dose-dependent contrast medium toxicity, including sodium and/or osmotic overload, in complex or prolonged procedures involving large total doses of contrast medium [2]. These dose-dependent effects are usually only relevant in vascular, rather than non-vascular, procedures.

THEORETICAL CONSIDERATIONS

In a typical intravenous urogram, 300 mg iodine/kg per patient might be used. For a 70 kg man, this represents 21 g iodine which could be provided, for example, by 50 ml of sodium iothalamate solution containing 420 mg I/ml. Several rules of thumb are in use to define an 'upper limit' for the total contrast medium dose. Typically, a maximum in a healthy adult of approximately 3 times the above level, i.e. 1000 mg I/kg, might be taken as a working figure. This may be compared with the lethal dose 50% (LD_{50}) measurement, for conventional contrast agents, of 8000 mg I/kg (a comparison of uncertain relevance to the clinical situation but the only type of guide available). There would appear to be some safety margin here, giving scope for higher doses to be used if the examination is vital and demands it. Concerning nephrotoxicity, Golman [13] would support a figure in this range as being an appropriate limit. There is clearly a question of clinical judgement in such decisions, to be exercised in the context of the individual patient, the importance of the procedure and the consequences of not proceeding with it. The time over which the total dose is to be administered is also clearly an important consideration, remembering that the half-life of contrast agents in the circulation is approximately 2 hours.

One other important factor now to be borne in mind is the availability of the better low osmolar non-ionic contrast agents. These have higher LD_{50}s, present a lower osmotic load and contain little or no sodium [9]. Whatever arbitrary upper limit in mg I/kg is set for the conventional agents, it seems to be the case that twice this dose of a new agent can be used. Thus, if 1000 mg I/kg may be given in the form of a conventional agent, possibly 2000 mg I/kg of a new agent may be given; if this is in the form of, say, Omnipaque 300 (iohexol, 300 mg I/ml; Nycomed, Oslo), this would mean a volume dose for a 70 kg man of $(2000 \times 70)/300 = 450$ ml. The non-ionic dimers offer an even greater margin of safety.

Even higher doses may perhaps be used if the contrast medium is administered over an extended period. Such high doses should never be given without serious thought, but if the procedure is vital, either for diagnosis or therapy, then they may be given with caution, detailed decisions being tailored to the individual patient. A reasonable precaution, with the risk to renal function in mind, is to make sure that the patient is well hydrated at the start of the procedure (see below).

DIGITAL SYSTEMS

The future of medical imaging is a digital one and a number of digital systems are already in widespread use, digital angiography being the most widely available [3]. The advantages for interventional radiology lie in the ability of such systems rapidly to acquire, store, retrieve and manipulate images, a great boon in long procedures. Another advantage is that, although digital systems have poorer spatial resolution than film, they have higher contrast resolution and sensitivity [3]. This makes it possible, and indeed necessary in some circumstances, to dilute contrast agents, thereby reducing the total doses used, There is little doubt that the combination of digital systems and less toxic contrast agents has made it possible to undertake many procedures which would otherwise have involved too much time and too much contrast medium [1–3].

ABNORMAL TREATMENT OF CONTRAST AGENTS

Mixing with other agents

Physical incompatibility is always to be feared and the mixing of drugs in a syringe is, in general, a bad principle. Nevertheless, on occasion, contrast agents are mixed with other drugs, if not actually deliberately in the syringe then incidentally, following inadequate flushing of the lines to clear one agent before another is injected. The best-known physical incompatibility is probably that of the vasodilator, papaverine, with ioxaglate [14]. Other incompatibilities have been reported, almost all being with ioxaglate and including cimetidine, protamine and benadryl [15].

The fact that physical incompatibilities with non-ionic contrast agents are virtually unknown should not, in the author's opinion, give the operator confidence to mix with them anything he or she chooses, and mixing should be avoided if at all possible.

Embolization mixtures frequently contain a combination of contrast agent, 50% dextrose or absolute alcohol and an antibiotic, but precipitation is not, of course, a great danger in embolization procedures!

Extreme conditions

A few radiologists have used boiling contrast material in embolization/sclerotherapy procedures. There is no danger to the integrity of the contrast agent in this. The compounds are all stable at their boiling points and

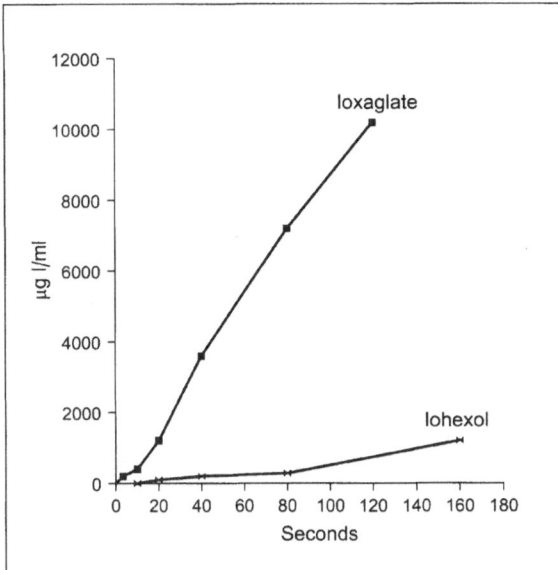

Figure 1 Liberation of inorganic iodide from dimeric and monomeric contrast agents (adapted from Laerum and Enge [17])

Lasers

Dawson et al. [16] and Laerum and Enge [17] have studied the effects on contrast agents of high-powered, hot-tip lasers. Contrast media are rapidly destroyed by the high energy densities reached and there is liberation of both particulate debris, available for distal embolization of the vessel [16], and inorganic iodide (Figure 1), which may have effects on thyroid uptake and cause systemic iodism. The monomeric contrast agents, ionic or non-ionic, are significantly more stable under these extreme conditions than the dimers, ionic or non-ionic.

The relevance of such apparently arcane studies lies in the fact that there may be a static column of contrast in the vessel undergoing a laser angioplasty which may be exposed to laser energy. Some now less frequently used varieties of hot-tip laser probes even allow the injection of contrast medium through their core during heating.

SUMMARY

The demands on water-soluble, iodinated radiocontrast agents are potentially much greater in interventional than in diagnostic radiological procedures, in that (a) significantly higher doses may be used, (b) there is a greater likelihood of encountering in vivo, if not in vitro, physical incompatibilities with drugs and materials, and (c) there is a greater likelihood that the agents will be subject to physical conditions for which they were not designed. These demands are currently best met by the non-ionic, monomeric, low osmolality agents. The non-ionic dimers may well replace the monomers in the near future.

REFERENCES

1. Dawson P. Iodinated intravenous contrast agents: a review. J Intervent Radiol. 1987; 2: 51–58.
2. Dawson P, Hemingway AP. Contrast agents doses in interventional radiology. J Intervent Radiol. 1987; 2: 145–146.
3. Dawson P. Digital subtraction angiography – a critical analysis. Clin Radiol. 1988; 39: 474–477.
4. Dawson P. Chemotoxicity of contrast media and clinical adverse effects: a review. Invest Radiol. 1985; 20: 52–59.
5. Dawson P. Factors dictating iodinated contrast agent toxicity. In: Katayama M, Brasch RC, eds. New dimensions in contrast media. Amsterdam: Excerpta Medica; 1991: 66–71.
6. Dawson P, Howell MJ. Pharmacology of the non-ionic dimers. Br J Radiol. 1986; 59: 987–991.
7. Dawson P. Cardiovascular effects of contrast agents. Am J Cardiol. 1989; 64: 2E–9E.
8. Grainger RG. The osmolality of intravascular contrast media. Br J Radiol. 1980; 53: 739–746.
9. Dawson P, Pitfield J. Hexabrix and the sodium problem. Br J Radiol. 1982; 55: 933–934.
10. Ansell G. Adverse reactions to contrast agents: scope of problem. Invest Radiol. 1970; 5: 374–384.
11. Katayama H. Adverse reactions to ionic and non-ionic contrast media. A report from the Japanese Committee on the Safety of Contrast Media. Radiology. 1990; 175: 621–628.
12. Dawson P, Sidhu P. Is there a role for corticosteroid prophylaxis in patients at increased risk of adverse reaction to intravascular contrast agents? Clin Radiol. 1993; 48: 225–226.
13. Golman K. Private communication, 1986.
14. Shah SJ, Gerlack AJ. Incompatibility of Hexabrix and papaverine and peripheral angiography. Radiology. 1987; 162: 619–620.
15. Irving MD, Burbridge BE. Incompatibility of contrast agents with intravascular medications. Radiology. 1989; 173: 91–92.
16. Dawson P, Booth A, Machan M. Laser angioplasty. Enough to make your blood boil. Br J Radiol. 1994;67:346–348.
17. Laerum F, Enge I. Contrast media properties in interventional radiology. In: Inge I, Edgren J, eds. Patient safety and adverse events in contrast media examinations. Amsterdam: Excerpta Medica; 1989: 147–158.

This paper was first published in *Advances in X-Ray Contrast*. 1994;2:18–21.

UPDATE

No significant new developments have taken place in this field but the notes on non-ionic monomers vs non-ionic dimers in platelet activation and degranulation appended to the article on 'The role of contrast agents in thromboembolic phenomena in clinical angiography' (pages 20–28 in this volume) may be relevant to such interventional procedures as general and coronary angioplasty.

P. Dawson and W. Clauss, (eds.), Advances in X-Ray Contrast: Collected Papers. 57–62
© *1998 Kluwer Academic Publishers.*

Contrast enhancement in computed tomography of the liver, pancreas and spleen

Andreas Adam, FRCP, FRCR
Guy's and St. Thomas' Hospitals, London SE1 9RT, UK

INTRODUCTION

Despite recent advances in magnetic resonance imaging (MRI), CT remains the investigation of choice for the examination of the liver, pancreas and spleen when ultrasound (US) has failed to provide a diagnosis. It provides an overall view of the upper abdomen and this is a great advantage, as diseases affecting one of these three organs frequently present with secondary abnormalities in the other two. For example, pancreatic carcinoma may be associated with hepatic and, occasionally, splenic metastases. Another example is cirrhosis of the liver due to chronic alcoholism, which may be associated with pancreatitis and splenic varices.

In most conditions in the liver, pancreas and spleen being investigated by CT scanning, the use of a contrast medium (CM) provides very useful additional information about the lesion itself, the normal parenchyma of the organ being examined and the blood vessels within that organ.

The exquisite contrast sensitivity of CT and its ability to measure X-ray attenuation accurately can provide diagnostic information, such as the occurrence of haemorrhage in a pancreatic pseudocyst, and can demonstrate minute amounts of calcification in the pancreas in cases of chronic pancreatitis.

A contrast-enhanced CT scan is usually indicated when an US study has not provided a diagnosis. MRI is challenging CT, especially in the investigation of liver diseases, but the speed and convenience of CT and the ease with which interventional procedures can be performed, combined with its more widespread availability, make it preferable to MRI in most centres.

More invasive studies, such as CT arteriography (CTA), CT-arterioportography (CTAP) and CT following hepatic intra-arterial Lipiodol (HIAL), are usually reserved for patients with primary or metastatic liver tumours being considered for partial hepatic resection. These studies are usually considered as the definitive investigations of such patients and are performed when other examinations have not demonstrated a lesion in the part of the liver which is to be preserved at surgery.

WHICH CONTRAST MEDIUM?

Dynamic CT scanning, CTA, CTAP and delayed CT scanning are performed using water-soluble, iodinated CM. Although small differences have been demonstrated in the rate of diffusion of CM into the hepatic parenchyma, the magnitude of these differences is not such as to affect the choice of CM in practice. In making the selection, the principles applied are those which govern the choice of CM for intravascular use in general.

The volume of CM administered during dynamic CT of the liver can be significant; in some centres it is common to use 180 ml of 60% CM [1]. It is well known that patients with normal cardiac function can tolerate an acute intravascular volume of 1L. The volume of 180 ml of 60% ionic CM is equivalent to 750 ml/L of normal saline; this volume load of CM has been accepted as safe in adequately hydrated patients with normal cardiorenal function who undergo angiographic procedures. The only difference between the method of CM delivery employed in angiography (3 ml/kg body wt/h) and the technique usually employed in CT is that the CM is delivered over 2 minutes with due regard for the patient's cardiac function. Non-ionic CM, which has approximately half the osmolality of ionic CM, can be used as an alternative for patients with abnormal cardiac function. In patients with normal baseline serum creatinine levels there is no abnormal elevation of serum creatinine at 24, 48 and 72 hours after the procedure. In patients with serum creatinine levels greater than 1.5 mg/dl (132.6 μmol/L), a non-contrast-enhanced CT scan should be obtained. If

58

results from the non-contrast-enhanced CT scan are negative in a patient with clinically suspected liver metastases, the use of MRI should be considered.

Lipiodol injected selectively into the hepatic artery is taken up by tumours in a variety of patterns. Normal hepatic parenchyma also takes up the Lipiodol, but the CM is cleared from normal liver within approximately 1 week, whereas it is retained in tumours. In general, vascular tumours such as hepatomas take up Lipiodol in a diffuse manner, whereas avascular lesions may not retain it at all or may demonstrate uptake only around the periphery of the lesion. It is thought that Lipiodol is taken up by tumours due to some abnormality of neoplastic vasculature which encourages leakage of CM into the tumour. Another explanation is that Kupffer cells clear Lipiodol from the normal hepatic parenchyma, but as such cells do not exist within neoplastic tissue, Lipiodol is retained within the latter. Usually, approximately 10 ml Lipiodol emulsion is injected into the hepatic artery and the CT scan is performed 7–10 days later but both the contrast volume and the timing of the examination vary considerably from centre to centre. In a study at Hammersmith Hospital, oily embolization of the liver as a diagnostic technique did not prove a reliable method of examination [2]. A study of 20 patients found that: (a) the technique was not risk-free, with complications in four patients, one of whom required emergency surgery for acute cholecystitis, (b) the Bruneton classification of patterns of Lipiodol uptake [3] was far too simplistic to be relied upon, (c) residual Lipiodol, particularly in the left lobe of the liver, was problematic, making it difficult to distinguish normal from diseased liver, and (d) the technique did not contribute to decisions about surgical management in any one of the 20 cases. This experience was disappointing and in marked contrast to enthusiastic reports by some other authors.

Promising results have been obtained in recent years with intravenously administered emulsified oily CM. These have been taken up by the liver parenchyma and focal lesions appear as low attenuation masses within the opacified hepatic parenchyma on CT. Several such agents have been tried over the years but most of them have proved too hepatotoxic for clinical use. A new agent, intraiodol, which was developed in Sweden, appears to be less toxic than previous agents but has

been used only in a relatively small number of patients to date, and firm conclusions cannot yet be drawn.

METHODS OF CT EXAMINATION USING WATER-SOLUBLE CONTRAST MEDIA

The spleen

Specific examinations of the spleen [4] by CT scanning are rarely carried out and the organ is usually inspected when a study of the liver is done, in which case, the volume of contrast and timing of scans are determined by the type of liver study being performed. Nevertheless, when splenic lesions are being sought specifically, it is best to administer approximately 150 ml of 60% CM, as described below for dynamic liver CT, but to delay scanning until 90–120 seconds after the beginning of the injection. This is because scans performed soon after the injection of CM are likely to show patchy areas of unequal attenuation due to differential flow patterns in the red and white splenic pulp. Later on, equalization of splenic parenchymal opacification increases the likelihood of lesions being detected and reduces the number of false-positive and false-negative results.

The pancreas

Dynamic CT scanning is the method of choice for routine examination of the pancreas [5]. Approximately 150 ml of CM can be injected in a biphasic technique: 50 ml are given at a rate of 5 ml/s for 10 seconds, followed by 1 ml/s for 100 seconds. Magnified contiguous sections of the pancreas are obtained every 4 or 5 seconds. It is best to use a dedicated CT volume flow rate injector that can be operated by a radiographer from the CT console. CM given by hand injection is not as accurate with respect to its timing, not as reproducible and not as convenient. With a volume flow rate injector, CM is delivered through standard venous cannulae (19 or 20 gauge) either 1¼ inches (3.2 cm) or 2 inches (5.1 cm) in length, preferably into antecubital veins. The radiologist must palpate the injection site to ensure that the CM delivered through the plastic venous cannula does not extravasate. If extravasation occurs, the injection should be stopped immediately. The optimal method

for delivering CM through antecubital veins is to position the patient's arm at a right-angle to the chest by placing the palm of the hand against the face of the CT gantry; this ensures that injected CM is not constricted at the thoracic outlet. Unlike the liver, which is mainly supplied by the portal vein, the pancreas has an arterial blood supply and the pancreatic parenchyma enhances earlier than the liver. Scanning should start 15–20 seconds after the beginning of the injection of CM. Almost all examinations will be completed in under 2 minutes, well within the period of excellent opacification of the pancreas.

Dynamic CT of the pancreas is a very accurate way of establishing the presence of vascular encasement by malignant tumours and demonstrating the extent of pancreatic neoplasms. In most centres, three or four scans are performed during held inspiration. The patient then takes another breath and the procedure is repeated until the whole organ has been scanned.

With the advent of faster CT scanners with tubes of greater heat capacity, it has become possible to scan the pancreas very fast while the patient breathes quietly. As a bolus, 100 ml CM is administered by hand and scanning begins immediately after the end of the injection. Scans obtained using this method may demonstrate some respiratory movement artefacts, but if a scanner capable of 2-s scans and a 3.5–5.5-s interscan delay is used, the artefact is minimal and does not affect the diagnostic quality of the examination, which is fully equivalent to that of CT scans obtained using the conventional technique described above. There are several advantages to this approach: a smaller volume of CM is used, the volume flow rate injector is unnecessary and the examination is completed in less time than with the conventional method – usually in less than 90 seconds. In addition, the fact that the patient is allowed to breath during the examination has obvious advantages with elderly patients and those with respiratory ailments who may find it difficult to suspend respiration.

Helical CT is being increasingly used in examining the pancreas [6]. In fact, many, including the author, would consider helical CT the method of choice for CT of this organ, as it provides high-quality images during the optimum period of parenchymal enhancement. Low-mAs scans are initially obtained to localize the gland before administering contrast material. Scanning can be done from head to foot or foot to head. Contrast material is injected at 2 ml/s for a total of 150 ml, and a 70-s delay between the start of the injection and the start of scanning is chosen. Because of the speed of helical imaging, thin (5 mm) sections of the pancreas may be obtained, with the entire liver also being evaluated during the non-equilibrium phase of the same injection of contrast material. Helical CT is very useful in detecting vascular invasion in patients with pancreatic carcinoma.

Very occasionally, CT arteriography is useful for the demonstration of vascular neoplasms of the pancreas. A catheter is inserted selectively into the coeliac axis and 60% CM is injected at 1 ml/s for 50 seconds. Dynamic scanning with incremental table movement is used to examine the whole organ. Scanning begins immediately after the start of the injection. Vascular lesions such as insulinomas are shown as hyperattenuating masses.

The liver

Dynamic hepatic CT

It is best to use a volume flow rate injector, as described above for pancreatic CT. It is also best to use a biphasic injection, beginning with a high flow rate; this causes a high peak contrast enhancement and delayed onset of equilibrium (and, therefore, longer optimal scanning interval). A volume of 50 ml is administered at 5 ml/s, followed by 100 ml at 2 ml/s. Scanning begins 30 seconds after the start of the injection [7]. Scanning with incremental table movement begins 40 seconds after the start of the injection and continues until the whole of the liver has been examined. The normal liver is supplied 75% from the portal vein and 25% from the hepatic artery, while metastases receive virtually 100% of their blood supply from the hepatic artery. Hepatic parenchymal enhancement reaches a plateau approximately 40 s after a bolus injection of CM. Most metastases are hypovascular and appear as low-attenuation lesions within the opacified parenchyma. Tumours that may be hypervascular in relation to normal hepatic parenchyma (e.g. primary hepatoma and metastases from pancreatic islet cell tumour, carcinoid, and renal cell carcinoma) may become isodense during the non-equilibrium phase of maximum hepatic enhancement [8]. Patients with suspected hypervascular tumours should have both a non-contrast and a dynamic post-contrast study.

CM delivered as described above ensures positive enhancement of the hepatic veins in addition to the portal veins, so that detected lesions can be located with respect to specific hepatic lobes and segments. Hepatic CT usually requires 12–20 contiguous sections (average, 16) and can be achieved in less than 2 minutes after the beginning of scanning in almost all patients, when using a modern fast CT scanner.

Unfortunately, in many centres, infusion techniques are still used in CT of the liver. With these methods, significant portions of the liver may not be examined until 5–10 minutes after the beginning of the CM infusion – this may result in metastases being isodense in the normal liver. In other cases, the increase in attenuation of the hepatic parenchyma may be insufficient to reveal small lesions. Dynamic CT of the liver, as described above, should be the routine method of scanning this organ [9,10]. If a volume flow rate injector is not available, hand injection of a bolus of CM followed by dynamic scanning is still better than an infusion technique.

Compared with unenhanced CT, use of dynamic, sequential hepatic CT does not markedly increase the number of patients correctly diagnosed as having liver metastases, but the number of lesions detected can be increased by as much as 40% and this is a most important consideration for patients being considered for partial hepatic resection.

Helical CT is an excellent way of dynamic CT scanning of the liver. A volume of 150 ml of CM is injected intravenously at a rate of 3 ml/s. Scanning begins 70 seconds after starting the injection. For routine evaluation of the liver, 7-mm collimation and a 30-s helical exposure at 330 mAs are used. The entire liver is covered in approximately 21–25 seconds. Scanning occurs well within the non-equilibrium phase, near the peak of hepatic enhancement, and before the decays in attenuation for the aorta and the liver become parallel during the equilibrium phase. Even if 5-mm collimation is used, the entire liver can be scanned in 30 seconds in most patients. Because helical data are continuous, the location in which slices are reconstructed can be selected retrospectively by the radiologist. For small lesions in the liver, it may be useful to reconstruct 10-mm 'thick' slices every 5 mm, to reduce partial volume averaging. Although the value of detecting small lesions is debatable, reconstructing overlapping slices should make small haemangiomas and metastases more apparent.

Delayed hepatic CT

Delayed hepatic CT [11,12] is a technique that uses CM contained within the interstitial spaces of the liver and within hepatocytes 4–6 hours after initial injection. This represents the small percentage of CM 'vicariously' taken up and excreted by the liver, as well as CM that remains in equilibrium with that circulating intravascularly. Provided that an adequate iodine load, at least 60 g, has been used initially, a 20-Hounsfield unit (HU) elevation of hepatic CT number is seen at 4–6 hours. The CM is administered intravenously and, as the examination is not performed until much later, the injection can be given quite slowly.

Delayed hepatic CT is a very sensitive technique in the detection of hepatic metastases and has a lower false-positive rate than CTAP. Nevertheless, few centres use this method routinely, mainly because it is inconvenient to schedule patients to be examined 4–6 hours after the initial injection of CM.

CT arteriography

Hepatic artery CTA is performed following selective hepatic arteriography. The arteriogram is obtained to define vascular anatomy for the surgeon, as well as to detect additional hepatic lesions. If the hepatic tumour appears to be resectable at the completion of arteriography, the catheter is left in the hepatic artery and the patient is transferred to the CT scanner. CTA is best carried out using a helical scanner. This minimizes the volume of contrast being used and usually allows the whole of the liver to be examined during a single breathhold. During scanning, 30% CM is infused through the hepatic artery at a rate of 1–2 ml/s. Approximately 70–100 ml CM are required for most patients. Thus, the additional iodine load to the patient is only minimal and is usually quite safe.

Hepatic artery CTA has been shown to be more sensitive than incremental dynamic CT for specific lesion detection. Approximately 30–55% of patients will have additional lesions detected.

A significant proportion of patients will have accessory hepatic arteries which must be catheterized, otherwise lesions supplied by those arteries will be missed. Metastases receive virtually all their blood supply from the hepatic artery, unlike the normal hepatic parenchyma, which is supplied by both hepatic artery and portal vein. CTA-identified metastases have a higher attenuation than the surrounding hepatic parenchyma and vascular lesions are easier to visualize on CTA than relatively avascular tumours.

One of the pitfalls of hepatic CTA is that layering or unusual flow patterns may result in the liver. These patterns correspond to main or subsegmental branches of the hepatic artery, which are more or less opacified. Hepatic contrast differences are primarily the result of flow going to one portion of the liver through the catheter, while the adjacent portion of the liver does not receive CM and thus is not opacified.

CT arterioportography

The basic idea behind CTAP is that normal liver parenchyma is enhanced by contrast medium delivered selectively via the superior mesenteric artery to the portal vein, whereas liver neoplasms receiving their blood supply mainly from the hepatic artery remain unenhanced during the portal and parenchymal phase of contrast material distribution [1]. A recent study established that optimum parenchymal enhancement is achieved between 18 ± 4 seconds and 67 ± 15 seconds after injection of contrast material into the superior mesenteric artery [13]. To scan the liver within such a narrow time-window, a helical CT technique is necessary. We use 150 ml of 60% contrast medium at 3 ml/s; the helical CT sequence is started 20 seconds after the beginning of the injection. In CTAP the injected CM is delivered selectively into the portal venous supply without distribution to, and dilution with, the central blood volume. This results in greater hepatic parenchymal enhancement and contrast differentiation between focal lesions and background. CTAP is easier to implement than hepatic artery injection CT because the catheter tip needs only to be placed in the superior mesenteric artery distal to any anomalous hepatic artery branches. Parenchymal enhancement of 80–100 HU can be achieved, compared with parenchymal enhancement of 50–70 HU achieved with intravenous bolus injection.

CTAP is an exquisitely sensitive technique for the detection of hepatic metastases [14–18]. Perfusion defects may be observed due to incomplete admixture of enhanced blood in the superior mesenteric vein with unenhanced blood in the splenic vein, resulting in hypoperfusion of the left hepatic lobe. In addition, central metastases may compress central portal vein branches, resulting in hypoperfusion defects. Although non-tumorous attenuation differences are significantly more frequent with CTAP than with dynamic CT, they are seldom a diagnostic problem because of their geographic pattern. In patients in whom it is unclear whether a hypoperfusion defect or a true focal lesion exists, it is advisable to perform a delayed hepatic CT study 4–6 hours after CTAP. However, lesions may be missed in areas which have not opacified sufficiently and it is important not to interpret CTAP in isolation from a conventional dynamic study and, if necessary, other examinations, such as US and MRI.

WHAT ARE THE COMPLICATIONS OF THESE PROCEDURES?

All methods of CT which utilize iodinated CM may be associated with hypersensitivity reactions, with disturbances of renal or cardiovascular function, with clotting disorders, pyrexia and, occasionally, with other rarer reactions. CT arteriography, CT arterioportography and Lipiodol CT may also be associated with the various complications of angiography.

Patients' tolerance of hepatic intra-arterial Lipiodol CT is usually excellent in cases of selective hepatic artery injection. Occasionally, non-selective injection into the coeliac axis is utilized and this sometimes results in certain side-effects immediately after the injection: approximately one-third of patients experience nausea or vomiting which regresses spontaneously within 15–20 minutes. In patients with an accessory hepatic artery arising from the superior mesenteric artery, an attempt to inject Lipiodol selectively into the accessory hepatic branch may result in a reflux of Lipiodol into the superior mesenteric artery. In such patients, diarrhoea may be observed for approximately 6 hours, but tends to resolve without sequelae. Acute cholecystitis requiring cholecystectomy has been described following hepatic artery injection of Lipiodol [2].

62

CONCLUSIONS

In the investigation of the liver, spleen and pancreas, dynamic CT following the intravenous injection of CM should follow US scanning when the latter has failed to provide a diagnosis. In patients being considered for partial hepatic resection, in whom dynamic CT has not revealed any lesions in the part of the liver which is to be preserved, CTAP is probably the investigation of choice. If this procedure reveals definite lesions, surgery is contraindicated. If very small lesions of questionable significance are detected, it is best to proceed to surgery and confirm the presence of such lesions with intra-operative US rather than deny the patient the chance of a cure.

Helical CT has made a major contribution in imaging abdominal organs [19] as it enables optimum utilization of contrast enhancement.

REFERENCES

1. Foley WD. Dynamic hepatic CT scanning. AJR. 1989; 152: 272–274.
2. Dawson P, Adam A, Banks L. Diagnostic iodized oil embolization of liver tumours – the Hammersmith experience. Eur J Radiol. 1993; 16: 201–206.
3. Bruneton J-N, Kerboul P, Grimaldi C. Hepatic intra-arterial Lipiodol technique, semiologic patterns and value of hepatic tumors. Gastrointest Radiol. 1988; 13: 45–51.
4. Federle MP. Computed tomography of the spleen. In: Moss AA, Gamsu G, Genant HK, eds. Computed tomography of the body. Philadelphia: WB Saunders Company. 1983 : 877–906.
5. Federle MP, Goldberg HI. Computed tomography of the pancreas. In: Moss AA, Gamsu G, Genant HK, eds. Computed tomography of the body. Philadelphia: WB Saunders Company. 1983: 699–762.
6. Ibukuro K, Charnsangavej C, Cinqualbre A, DuBrow RA, Varma GK, Wallace S. Helical CT and multiplanar reformation of the pancreas and peripancreatic vessels. Radiology. 1993; 189(P): 256–257.
7. Heiken JP, Brink JA, McClennan BL, Sagel SS, Forman HP, DiCroce J. Dynamic contrast-enchanced CT of the liver: comparison of contrast medium injection rates and uniphasic and biphasic injection protocols. Radiology. 1993; 187: 327–331.
8. Bressler El, Alpern MB, Glazer GM, Francis IR, Esminger WD. Hypervascular hepatic metastases: CT evaluation. Radiology. 1987; 162: 49–51.
9. DuBrow RA, David CL, Libshitz HI, Lorigan JG. Detection of hepatic metastases in breast cancer: the role of nonenhanced and enhanced CT scanning. J Comput Assist Tomogr. 1990; 148: 366–369.
10. Patten RM, Byun J-Y, Freeny PC. CT of hypervascular hepatic tumours: are unenhanced scans necessary for diagnosis? AJR. 1993; 161: 979–984.
11. Nelson RC, Chezmar JL, Sugarbaker PH, Bernadino ME. Hepatic tumors: comparison of CT during arterial portography, delayed CT, and MR imaging for preoperative evaluation. Radiology. 1989; 172: 27–34.
12. Heiken JP, Weyman PJ, Lee JLT, et al. Detection of focal hepatic masses: prospective evaluation with CT, delayed CT, CT during arterial portography, and MR imaging. Radiology. 1989; 171: 47–51.
13. Graf O, Dock WI, Lammer J, et al. Determination of optimal time window for liver scanning with CT during arterial portography. Radiology. 1994; 190: 43–47.
14. Matsui O, Takashima T, Kodoya M, et al. Liver metastases from colorectal cancers: detection with CT during arterial portography. Radiology. 1987; 165: 65–69.
15. Nelson RC, Chezmar JL, Sugarbaker PH, Murray DR, Bernadino ME. Preoperative localization of focal liver lesions to specific liver segments: utility of CT during arterial portography. Radiology. 1990; 178: 89–94.
16. Sayer P, Roche A, Gad M, et al. Preoperative segmental localization of hepatic masses: utility of three-dimensional CT during arterial portography. Radiology. 1991; 180: 653–658.
17. Sayer P, Levesque M, Elias D, Zeitoun G, Roche A. Preoperative assessment of resectability of hepatic metastases from colonic carcinoma: CT portography vs sonography and dynamic CT. AJR. 1992; 159: 741–744.
18. Sayer P, Levesque M, Elias D, Zeitoun G, Roche A. Detection of liver metastases from colorectal cancer: comparison of intraoperative US and CT during arterial portography. Radiology. 1992; 183: 541–544.
19. Zeman RK, Fox SH, Silverman PM, et al. Helical (spiral) CT of the abdomen. AJR. 1993; 160: 719–725.

This paper was first published in *Advances in X-Ray Contrast*. 1995;2:34–39.

P. Dawson and W. Clauss, (eds.), Advances in X-Ray Contrast: Collected Papers. 63–66
© *1998 Kluwer Academic Publishers.*

Spiral computed tomography – a short overview

Mathias Langer, MD, FICA
Radiologische Universitätsklinik der Albert-Ludwigs-Universität Freiburg, D-79106 Freiburg, Germany

INTRODUCTION

During the late 1980s and at the beginning of the 90s, fast CT (computed tomography) scanners with an examination time of less than 1 second, based on the slipring technology, were introduced into clinical practice. These scanners offered the opportunity to have continuous data acquisition over a multitude of 360° scans, and it seemed feasible that volume imaging could be obtained during one breathhold by the patient. This eliminated one of the disadvantages of conventional examinations, namely, the risk of patient motion between two scans.

Spiral CT (or helical scanning) was first described in 1989 by Kalender et al. [1]. Immediately after its implementation in the early 90s, multiple applications from thoracic to abdominal imaging were described and integrated into clinical practice [2–4].

The essential feature of spiral CT is the fact that, during a continuous rotation of the X-ray tube and simultaneous continuous data acquisition, the patient is slowly moved in a craniocaudal or caudocranial fashion through the X-ray beam. When this new technique was introduced into routine clinical practice, thoracic studies were the first target. It was clear that the variation of inspiration or expiration depth could easily be the reason of a reduction in detection rate of focal pulmonary masses, and a detailed analysis by Remy-Jardin et al. [5] showed that, in comparison with conventional CT scanning, helical scanning improved the detection rate of pulmonary metastases by a factor of 2–3. Subsequently, pulmonary spiral CT has become the gold standard for pulmonary CT (Figure 1).

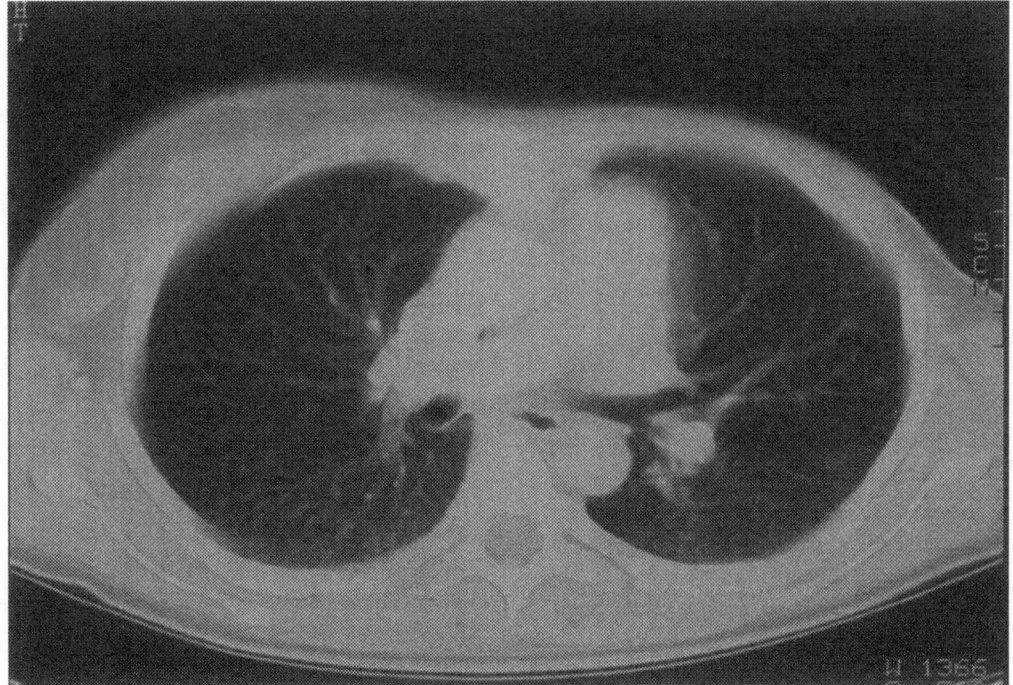

Figure 1 Spiral CT of the lung demonstrating pulmonary vascular architecture as well as disseminated small (< 1 cm) pulmonary metastases

AORTIC EXAMINATION

For examination of the thoracic aorta, it was clear that an examination time of less than 30 seconds would be ideal to provide homogeneous and good contrast within the aortic lumen for the detection of aortic dissections (Figure 2).

The application rates, and especially the application volume, for spiral CT of the thorax had to be modified according to the increased examination speed. It could be shown by Prokop et al. [6] and other groups of investigators [7–9] that spiral CT has the possibility not only to detect the flap in an aortic dissection and to differentiate between type A and B dissections, but also to demonstrate clearly the entry and re-entry points. In addition to an excellent demonstration of the abdominal aorta, it could be seen by these groups that the involvement of mesenteric arteries in the dissecting process or their occlusion by a thrombosed false lumen was easily imaged.

The image quality of 2- and 3-D reconstructions based on a spiral CT dataset was the basis for the integration of maximum intensity projection algorithms as an angiographic tool in CT. The maximum intensity projection protocol which was developed at Stanford University Medical Center proved to be a new method for evaluating arterial diseases in CT. This maximum intensity projection protocol and a 3-D technique have been applied in the past, not only to the investigation of aortic dissections and aortic aneurysms, but also to vascular occlusive disease in the superaortic vessels, the abdominal and pelvic aorta [10,11], and especially for renal arterial stenoses. In a very interesting analysis by Oates [9] of efficiency and cost factors for spiral CT angiography (CTA) in comparison with conventional angiography, the author found that CTA, a less invasive diagnostic tool, yielded the same information in 30 minutes to 1 hour, on an outpatient basis. A conventional aortogram for abdominal aortic aneurysms needed 24 hours, when angiography was performed on an inpatient basis. The average cost of spiral CT in the Deaconess Hospital in Boston compared favourably with that of conventional angiography, both performed on an outpatient basis ($980 and $3000, respectively).

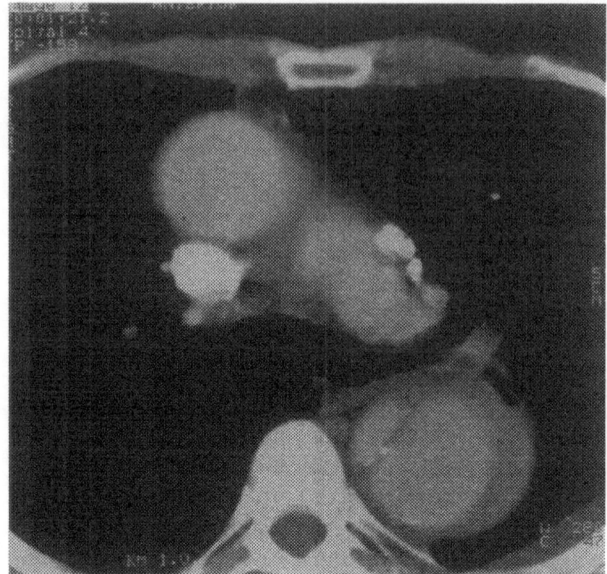

Figure 2 Contrast-enhanced spiral CT of the thorax, clearly demonstrating dissection of the descending aorta with periaortic haematoma, indicating partial rupture of the aortic dissection

LIVER

Spiral CT of the liver has taken some time to be accepted in clinical practice, due especially to the fact that calculating time for the spiral-CT images did not allow an evaluation of the liver with one contrast medium application in arterial and portal venous phases (Figure 3).

This has been overcome with the latest generation scanners, which offer the possibility of so-called 'shuttle' scanning, in which the first examination is done in one direction and the second follows immediately or at an operator-defined interval, in the opposite direction.

It has been demonstrated by Bluemke and Fishman [12], as well as by other groups [13], that detection and characterization of hepatic tumours are best carried out by rapid breathhold scanning of the liver in arterial and portal venous phases. This means that both perfusion states of the tumour are imaged in one examination with only one application of contrast material.

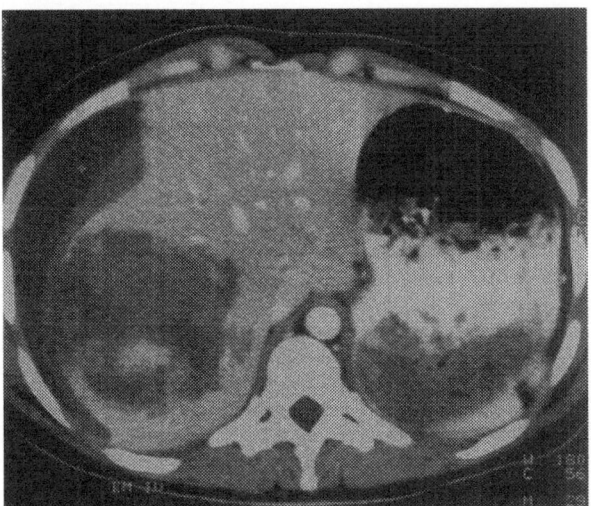

Figure 3 Spiral CT of the abdomen demonstrating perihepatic and subcapsular haemorrhage as well as large intrahepatic haematoma due to a bleeding hepatic adenoma, in a young female. Examination in the portal venous phase clearly shows vascular architecture of the liver as well as a huge intrahepatic bleeding, easily identified by the hyperdensity within the intrahepatic haematoma

It is clear that for all spiral CT studies with contrast material injection, especially for abdominal studies, a power-injector with exact triggering of the bolus for the beginning of the scan is mandatory [14]. Until now, this has been effected by a so-called test bolus and interval scanning at the same level. In the latest generation scanners, an automatic or semi-automatic triggering device will probably be incorporated, which will allow scanning to commence at the optimum increase in density at a given point of interest.

For hepatocellular carcinoma, the comparison of arterial and portal venous phases eliminates some differential diagnostic problems; it is feasible that the same applies to diagnosing focal nodular hyperplasias and adenomas. Metastatic liver disease will probably be imaged in a quality which will equal that of the more invasive technique of spiral CT arterioportography, with contrast medium injection into the splenic or mesenteric artery. Intraindividual, prospective, blinded studies are still the object of clinical research and are not yet published.

CONTRAST MEDIA

The application of contrast material has to be redefined, and the examination protocols have to be adjusted to the new technology. In different studies it could be shown that the volume necessary for a spiral CT examination could be reduced by a factor of 2–3 compared with conventional studies. For the purpose of standardizing the technology, a consensus meeting was held in February 1995 in Berlin, Germany, organized by the Universities of Regensburg and Marburg. A review of this meeting will be published in a future issue of *Advances in X-Ray Contrast*. In summarizing the data, it was clearly demonstrated, and an overall consensus was established, that spiral CT with intravenous contrast medium application is a suitable method for the examination of the head and neck region, the thorax and the whole abdomen. Spiral CT of these regions, performed in an adequate manner, will be the method to be compared with MRI examinations, with respect to sensitivity, specificity and accuracy, as well as the side-effects of both technologies and their respective contrast media.

TRAUMA

CT techniques, especially those using fast scanning methods, are essential for the evaluation of a trauma patient. Safe and accurate radiographic and CT diagnosis is essential for the evaluation and treatment of spinal trauma. Spinal osseous pathology has been one of the main applications of 2- and 3-D reconstructions in CT. Routinely, CT was primarily performed with contiguous, non-overlapping slices, and secondary 2- and 3-D reconstructions were performed [15,16]. Based on the spiral CT technology, there has been a considerable increase in quality of multiplanar reconstructions. Due to the fact that there is no motion artefact between the different calculated slices, and that there is no problem in calculating thin axial slices, reconstruction of 2- and 3-D CT sections can be performed with an excellent image quality. It has been demonstrated in multiple studies that the 2-D reconstructions particularly are of utmost importance for the analysis of vertebral pathology in a trauma patient. Together with the excellent demonstration of a

66

known pathological process by 3-D images, nearly all clinical questions and all operation planning can be based on these CT examinations.

It is even more interesting and essential for the reconstruction of complex maxillofacial injuries to have a complete overview of osseous pathology. In a prospective study, Buitrago-Téllez et al. [15] evaluated 2-D and 3-D reconstructed CT examinations in patients with a severe maxillofacial injury. It could be shown that bone fractures with dislocations larger than 2 mm could be excellently visualized by the CT examinations. This high image quality was essentially achieved by using spiral CT protocols and calculated image thicknesses of 1–2 mm. The study, as well as the results of other groups, shows that multiplanar CT examinations, based on a 3-D dataset acquired by spiral CT, are today the basis for diagnostics and treatment planning for patients with traumatic bone lesions in the maxillofacial region.

CONCLUSIONS

In summarizing the different indications for spiral CT briefly mentioned above, it can be stated that the implementation of this technology has brought CT a big step forward in better diagnosis of pathological processes, and in reducing costs by reducing examination time and contrast material volume, as well as by pushing the diagnostic potential of CT to new frontiers.

REFERENCES

1. Kalender WA, Seissler W, Vock P. Single-breath-hold spiral volumetric CT by continuous patient translation and scanner rotation. Radiology. 1989; 173(P): 414.
2. Kalender WA, Polacin A, Süss C. A comparison of conventional and spiral CT: an experimental study on the detection of spherical lesions. J Comput Assist Tomogr. 1994; 18: 167–176.
3. Kalender WA, Seissler W, Klotz E, Vock P. Spiral volumetric CT with single-breath-hold technique, continuous transport, and continuous scanner rotation. Radiology. 1990; 173: 181–183.
4. King K, Crawford CR. CT scanning with simultaneous patient translation. Radiology. 1990; 177(P): 108.
5. Remy-Jardin M, Remy J, Giraud F, Marquette C-H. Pulmonary nodules: detection with thick-section spiral CT versus conventional CT. Radiology. 1993; 187: 513–520.
6. Prokop M, Schaefer C, Kalender WA, Polacin A, Galanski M. Gefäßdarstellungen mit Spiral CT. Sonderdruck aus Radiologie. 1993; 33: 12.
7. Costello P, Gaa J. Spiral CT angiography of the abdominal aorta and its branches. Eur Radiol. 1993; 3: 359–365.
8. Galanski M, Prokop M, Chavan A, Schaefer CM, Jandeleit K, Nischelsky JE. Renal arterial stenoses: spiral CT angiography. Radiology. 1993; 189: 185–192.
9. Oates M. Spiral computed tomography angiography vs. conventional angiography. Efficiency and cost factors. Admin Radiol J. 1305 W. Glenoaks Blvd., Glendale, CA 91201, USA.
10. Ferstl FJ, Uhrmeister P, Flügel P, Blum U, Barke A, Landes G. Langer M. CT angiography with maximum intensity projection in the assessment of pelvic arterial disease. In: Pokieser H, Lechner G, eds., Advances in CT III. Vienna: Springer-Verlag; 1994.
11. Semba C, Dake M. New developments in vascular imaging. Spiral CT angiography. Admin Radiol J. 1305 W. Glenoaks Blvd., Glendale, CA 91201, USA.
12. Bluemke DA, Fishman EK. Spiral CT of the liver. AJR. 1993; 160: 787–792.
13. Oudkerk M, van Ooijen B, Mali SPM, Tjiam SL, Schmitz PIM, Wiggers T. Liver metastases from colorectal carcinoma: detection with continuous CT angiography. Radiology. 1992; 185: 157–161.
14. Ehritt-Braun C, Ferstl FJ, Burger D, Langer M. Optimisation and adaptation of intravenous administration of contrast medium in spiral volumetric CT. In: Pokieser H, Lechner G, eds., Advances in CT III. Vienna: Springer-Verlag; 1994.
15. Buitrago-Téllez CH, Wächter R, Ferstl F, Stoll P, Düker J, Langer M. 3-D-CT zur Befunddemonstration bei komplexen Gesichtsschädelverletzungen. Fortschr Röntgenstr. 1994; 160: 106–112.
16. Saeed M, Buitrago-Téllez H, Ferstl F, Boos S, Wimmer B, Langer M. Three-dimensional CT in the diagnosis of spinal trauma: comparison with plain film and two-dimensional CT examinations. Eur Radiol. 1994; 4: 161–166.

This paper was first published in *Advances in X-Ray Contrast*. 1995;2:50–53.

P. Dawson and W. Clauss, (eds.), Advances in X-Ray Contrast: Collected Papers. 67–69.

Electron beam computed tomography (EBCT)

This review of EBCT is in five parts. Parts 1 and 2 were first published in Advances in X-ray Contrast. 1995;2:54–56 and 57–60.

1. Technical aspects in EBCT

R Knapp, I Bangerl, D zur Nedden
Universitätsklinik für Radiodiagnostik, Radiologie II, Anichstrasse 35, A-6020 Innsbruck, Austria

INTRODUCTION

The EBCT scanner is a special type of computed tomography (CT) unit (Figure 1). It was first introduced into clinical practice in 1983 [1] and, like a 'conventional' CT machine, creates transaxial slices of the human body or skull.

The main difference between CT units and the EBCT scanner is the X-ray source. Whereas conventional CT machines use an X-ray tube rotating around the patient's body, exposing it to an X-ray beam which is attenuated by the object before it hits an array of detectors, the EBCT's X-ray source is an electron beam focused onto a fixed tungsten target. The electron beam is produced by an electron beam gun which is situated at the back of the machine. After it has been generated, the electron beam is focused on four tungsten targets, named after the first four letters of the alphabet (ABCD), that cover 210° of the gantry and are situated beneath the patient [2,3]. This device needs a large vacuum chamber in which an ultra-high vacuum of 10^{-7} torr is maintained. When the electron beam hits and sweeps the target, X-rays are released in a 30° fan-like beam, perpendicular to the patient's z-axis. This X-ray beam penetrates the object inside the gantry. The attenuated X-ray beam is measured by means of a fixed double array of detectors that cover 210° of the gantry on the opposite side of the targets. Detector segment 1 comprises 432 detectors; segment 2 comprises 864 detectors. In high-resolution mode, only segment 2 is used.

The X-ray energy is produced at 120 kV and 680 mA/s. These parameters cannot be changed, so X-ray energy is always proportional to scan time. One sweep along one target ring lasts for 50 or 100 ms. Switching time from one sweep to the next is 8 or 16 ms, respectively, on the same or another target. Slice thicknesses available are 1.5, 3, 6, 8 or 10 mm. Besides normal processing algorithms, edge-enhancing techniques for narrow collimation are possible.

OPERATING MODES

Single-slice mode

In this mode, the EBCT scanner operates similarly to a conventional CT unit. The patient is scanned slice by slice, with table incrementation after every slice. For this mode only, the C-target produces X-rays with a minimum scan time of 100 ms and a maximum scan time of 1.9 s. In the 100 ms mode, ECG-triggering of the machine is available at 0, 40 or 80% of the R–R interval. All slice thicknesses are available in this mode, and, as only one target ring is used in this mode, the machine produces only one image per X-ray exposition. After every image, the table is incremented as in a conventional X-ray machine.

Multi-slice mode

In this mode, scan time is 50 ms and switching time 8 ms, in general. Beam collimation is 8 mm, and no table incrementation is necessary until all targets have been used. From one to all four targets can be used and every

68

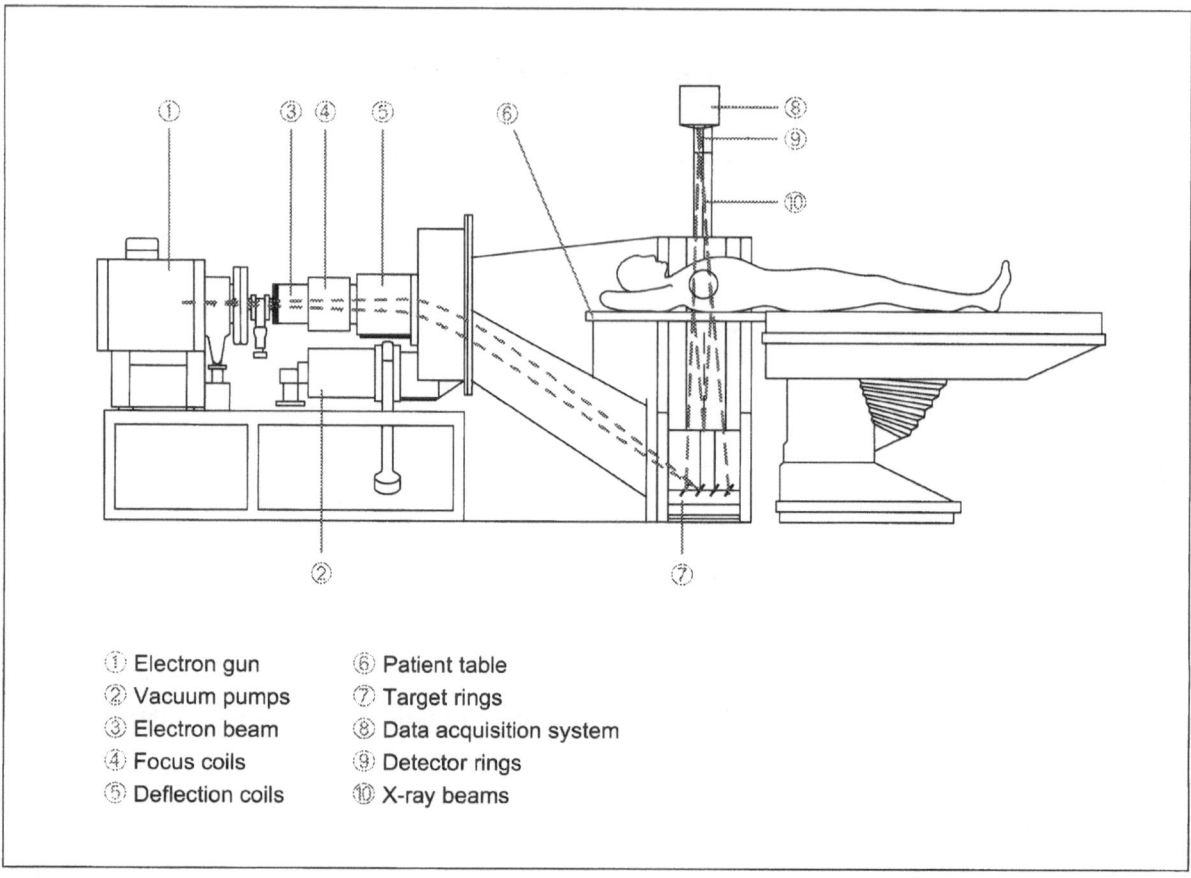

Figure 1 The EBCT scanner

① Electron gun	⑥ Patient table
② Vacuum pumps	⑦ Target rings
③ Electron beam	⑧ Data acquisition system
④ Focus coils	⑨ Detector rings
⑤ Deflection coils	⑩ X-ray beams

target produces two images, due to the two detector arrays. In this mode, all four targets can fire X-rays one after the other, and in 224 ms, eight images, covering 8 cm of the z-axis, are generated. The trade-off of this algorithm is that, after every pair of images produced by the double detector rings, a gap of 4 mm appears before the volume of interest is covered by the next pair of images. The number of scans available is limited to the memory of the computer (to date, 80 images). It is therefore possible to use fewer than all four detector rings and create more consecutive images per volume, to cover a longer period of time. On the other hand, one can use 'more' than the four targets when the table is incremented after all four targets have been used, and a larger volume along the z-axis of the patient can be imaged.

Flow mode

If the multi-slice mode is repeated every, or every other, heartbeat, triggered by the R-wave of the ECG, flow studies of the passage of a bolus of contrast medium throughout the whole heart and the great vessels can be generated. The inflow of the bolus of contrast medium can be visualized, followed by the peak and washout phase of the contrast agent. In this so-called 'flow-mode', every target is swept sequentially.

Cine mode

Cineloops can be achieved by producing images every 58 ms on the same level throughout one cardiac cycle. This means that every target ring fires X-rays at a frequency of 17 runs per second before the next target ring is used. For this purpose, the examination couch can be tilted in two directions to achieve views of the long and the short axis of the heart, so it is possible to generate near-real-time moving images of the heart and the great vessels. Because of this, a plateau of contrast enhancement of flowing blood should be achieved, to enhance the ventricular cavities and the blood inside the great vessels.

Continuous-volume mode

As with the spiral or helical modes of CT machines, the EBCT scanner can operate during continuous table incrementation. Slice thickness is always equivalent to table progress in the same time interval. This algorithm produces a volume dataset with a pitch of 1. Continuous scanning is possible for up to 20 seconds in one run.

PRACTICAL APPROACH

The main advantage of the EBCT scanner is its capability to produce much more X-ray energy in a much smaller time and for a longer period of time than in conventional X-ray tubes. This X-ray energy is delivered around the body to create an image without any moving parts inside the gantry, so any image degradation caused by motion, from either heart or vessel movement, can be avoided. This enables the scanner to image the heart and the great vessels without image blurring due to motion artifacts.

Critically ill patients who are not capable of holding their breath can be studied without any artifact arising from breathing. Anaesthesia is not needed for babies or small children. Physiological or pathophysiological movements, such as breathing or the heartbeat, can be studied with almost real-time resolution, in moving images.

As the image acquisition time in the EBCT scanning protocol is very short, the X-ray dose delivered to the patient is also small. This fact may cause noisy images because of the low milliamperage per slice. If temporal resolution is not the primary goal in a specific CT study, the signal-to-noise ratio can be improved by choosing repetitive 100 ms sweeps over one target, using the single-slice mode. For example, four sweeps in a thorax study, which means an acquisition time of 464 ms and delivers a dose of 272 mA per slice, or an abdominal study in which a dose of 544 mA is delivered in a scan time per slice of 928 ms. Theoretically, the dose can be advanced up to 1156 mA per slice when the 2-second algorithm is used. In the field of body imaging, the high level of available X-ray energy per slice or volume gives excellent image quality because of a low signal-to-noise ratio. This is also true if the conventional single-slice mode is used, as the volume-acquiring continuous volume scan is performed.

CONCLUSION

Compared with state-of-the-art spiral CT scanners, the acquisition time for an image is shortened by a factor of ten. This enables the EBCT scanner to be triggered by ECG and operate inside an R–R interval. However, this ultrafast EBCT scanner requires special care to be taken with the contrast protocols used. The synchronization of individual biological parameters, including contrast dilution after intravenous application and circulation time, is mandatory to achieve good results and benefit from contrast-saving fast protocols in EBCT.

REFERENCES

1. Boyd DP, Lipton MJ. Cardiac computed tomography. Proc IEEE. 1983; 71: 298–307.
2. Stanford W, Galvin JR, Weis RM, et al. Ultrafast computed tomography in cardiac imaging. Semin Ultrasound CT MR. 1991; 12: 45–60.
3. McCollough CH, Morin RL. The technical design and performance of ultrafast computed tomography. Radiol Clin North Am. 1994; 32: 521–536.

P. Dawson and W. Clauss, (eds.), Advances in X-Ray Contrast: Collected Papers. 70–73
© 1998 Kluwer Academic Publishers.

2. Protocols for the application of contrast media in EBCT

R Knapp, I Bangerl, D zur Nedden
Universitätsklinik für Radiodiagnostik, Radiologie II, Anichstrasse 35, A-6020 Innsbruck, Austria

INTRODUCTION

The protocols for the application of contrast media used in conventional or spiral CT scanners have to be modified in the case of the EBCT scanner, because of the superior scanning speed. For some special scanning protocols in EBCT, the method of application of contrast medium has to be completely different. The contrast medium protocol has to be linked to the scanning parameters in such a way as to fulfil the following criteria:

- There must be sufficient opacification of the vessels, organs or pathological structures of interest to distinguish them from the adjacent tissue.
- A minimum amount of contrast agent must be applied, in order to minimize the side-effects of the drug and reduce costs.
- A safe and easy method of application must be guaranteed.
- The protocol must cause a minimum feeling of discomfort to the patient.
- There must be a standardization of the contrast media protocol, in order to achieve comparable results in different patients.

To meet all of the above criteria may be very time-consuming during daily routine work. In this section, we would therefore like to suggest how to find the best way of achieving good results with minimum amounts of time and money.

TYPES OF CONTRAST AGENTS

Non-ionic contrast agents are generally used in CT. The minimal side-effects of non-ionic contrast agents are much better tolerated by the patients. Artifacts due to patient movement caused by nausea are minimized, so the number of poor quality scans, caused by contrast agent side-effects, is near zero.

IODINE CONTENT

Most CT examinations require a good opacification of arteries. As contrast agent is usually applied via an antecubital vein, the drug undergoes an extensive dilution process while passing through the circulation of the right heart and the lungs, before it reaches the organ of interest via its nutritional artery. The degree of contrast which is caused by the iodine content of the contrast agent is therefore diminished. For these applications, a contrast agent with the highest iodine concentration available is the best, because the volume applied can then be reduced.

One disadvantage of this technique can be artifacts caused by maximum density in the brachiocephalic veins and the superior vena cava, although this side-effect is rare if not more than 2.5 to 3 ml/s of contrast medium are given via an antecubital vein.

TOXICITY

The toxicity of the non-ionic contrast agents is presumably low. The dose of contrast agent is reduced when the serum creatinine level exceeds 2.5 mg %. In cases of impaired renal function, special surveillance of the patient after the application of contrast agent is mandatory. Hyperthyrotic patients, or those with myeloma, should be given contrast agent only if the examination is clinically unavoidable. Special pre-medication and diuresis are important, to guard against renal failure or thyrotoxic crisis. In such cases, there should be close co-operation with the clinicians involved, in order to facilitate the application of contrast medium. Patients with known contrast agent-induced allergic reactions are excluded from contrast media studies.

METHOD OF APPLICATION

A syringe-type injector is the best tool to guarantee a constant, non-pulsatile application of contrast medium. This is preferable to a roller-pump type of injector that produces a pulsatile flow at a higher flow rate. The pulsatile flow, especially at higher flow rates, can damage the wall of the venous injection site.

We use only monophasic contrast protocols, because most of the scanning is done during a few heartbeats. Most of the scanning protocols are therefore terminated before a plateau-like intravascular contrast curve is attained.

CIRCULATION TIME

Knowledge of the circulation time is mandatory for special applications in EBCT. Three methods of measurement of circulation time are in clinical use:

1. IV application of cardiogreen dye and measurement of the time the dye takes to arrive at the earlobe, using a photosensitometer.

2. IV application of magnesium sulphate, which causes a feeling of discomfort inside the throat after arrival in the arteries of the neck.

3. IV application of contrast medium and measurement of the density at the level and in the vessel of interest.

The circulation time is measured by administering a small bolus of contrast medium over a time period of 3 seconds, applied at the same flow rate to be used during the examination. For example, if a bolus of 80 ml is planned to be injected at a rate of 4 ml/s, 12 ml of contrast medium is given at a rate of 4 ml/s in order to estimate the circulation time. Scanning is started after a delay of 5 seconds at the region of interest and performed every other heartbeat. On a time-density curve, with the region of interest situated in the vessel under examination, the arrival of the bolus can clearly be seen by the increasing density in the region of interest. The circulation time in seconds is calculated as 5 + the arrival time of the bolus.

The advantage of this method is that the circulation time is known exactly at the level of interest of the arterial system. The cardiogreen dye method only estimates the circulation time, derived from the time the dye takes to arrive at the earlobe – the same disadvantage is found with the magnesium sulphate method. Another disadvantage of the magnesium sulphate method is that it is dependent on the co-operation of the patient. In most routine cases, especially the survey cases, a measurement of the circulation time is not necessary.

EBCT SCANNING PROTOCOLS

We would like to present our method of contrast agent application for the most important scanning protocols, as follows:

Thoracic survey

The examination of the thorax should be done in the timed, single-slice mode with a scan time per slice of 464 ms, which comprises a dose of 270 mA per slice. Scanning can be done continuously, one slice after the other in slabs. The length of the slabs along the z-axis of the patient is dependent on the ability of the patient to hold his or her breath. Scanning can be stopped at any time to give the patient the opportunity to breathe, and started again with the examination during a new period of apnoea. The examination can be completed within 2 to 6 breathing pauses.

Proper opacification of the supra-aortic vessels is mandatory at the beginning of the examination. This is guaranteed if 50 to 60 ml of contrast medium are administered during the first two periods of circulation time. Circulation time can be estimated to be 15 to 20 seconds, so if scanning commences after 30 seconds and contrast medium is applied at a rate of 2 ml/s, the criteria are fulfilled. The 2 ml/s protocol should be maintained until 15 seconds before the examination is terminated, so the last slices still have a good opacification of the vessels.

Heart (flow study)

A multi-slice flow study is carried out to determine how a bolus of contrast medium is driven through the heart [1]; the tungsten targets are swept sequentially, one after the other every other heartbeat, over a period

72

of twenty heartbeats. Circulation time is calculated as the bolus arrival time inside the right ventricle, according to the test bolus method described above.

As the bolus arrival time to the right side of the heart is rather short (usually about 10 seconds or less), a flow rate of 4 to 6 ml/s is necessary to provide sufficient volume of contrast medium for a good opacification of the right ventricle. The bolus should be terminated immediately after the scanning begins, to catch the washout phase of contrast medium in the right ventricle and see the inflow of contrast medium into the left ventricle.

As the end of the bolus is usually not as clearly defined as the bolus peak, an infusion of saline immediately following the bolus can improve the result of the study. The high flow rates used in this protocol can cause streak artifacts along the course of the superior vena cava. This can be overcome by injecting contrast medium via a pigtail catheter placed into the inferior vena cava just below the diaphragm. The clinical impact of this study is to be able to see a right-to-left shunt. The same protocol is used to prove the patency of aorto-coronary bypass grafts. Circulation time must be measured in the aortic root and scanning must be started later, to generate time–density curves over the bypass grafts which can be compared with time–density curves measured in the aorta.

For the calculation of myocardial perfusion, delayed images are necessary, to observe the equilibrium phase of contrast.

Heart (volume study)

The multi-slice mode of the scanner is also capable of generating cineloops of the beating heart. This is achieved when every target is used repetitively for 10 scans, starting at 0% of the R–R interval. In this protocol, two levels are imaged during one heartbeat.

To create cine modes of the beating heart, a good opacification of the blood pool of the atria and ventricles is mandatory. This is obtained by injecting contrast medium via an antecubital vein, using 2 ml of flow and a delay of 30 seconds, so measurement of the circulation time is not absolutely mandatory. As the protocol only lasts for eight heartbeats until scanning is terminated, only a small amount of contrast medium is needed and the contrast injection is terminated

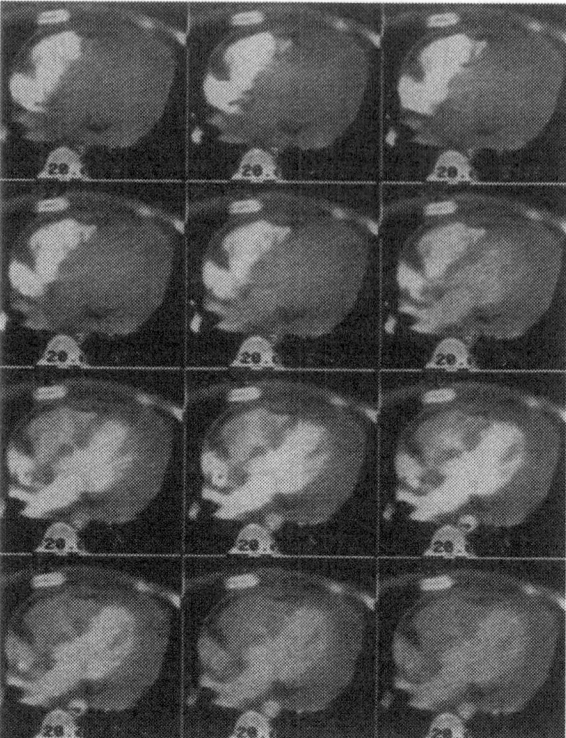

Figure 1 Sequence of a flow-mode study. The bolus passes through the heart

immediately after the scanning has begun. When all the images are added together, a cine loop of the beating heart can be created, with an almost real-time resolution (Figure 1) [2].

Abdominal survey

For the abdominal survey study, we use the same single-slice-timed algorithm as in the thorax survey, despite the higher amount of milliamperage needed to avoid noisy images. Scan time per slice, therefore, rises to 696 ms, to give 408 mA per slice. Depending on the major area of interest in the upper abdomen, the scan can be done in the arterial phase or in an equilibrium of arterial and portal venous phase. If the inferior vena cava below the renal veins is of interest, an extra drip infusion of contrast medium should be applied via a frontal foot vein.

Liver

Focal liver lesions may be differentiated by their different blood supply. Metastases almost always have a single arterial blood supply and no portovenous blood supply. This is best demonstrated by CT studies in which contrast medium is applied to the superior mesenteric artery, when contrast is transported solely by the portal vein inside the liver. Metastases appear hypodense during this CTAP procedure which is the most sensitive way to identify metastases inside the liver. Other lesions, such as focal nodular hyperplasia (FNH), have a flush-like early arterial enhancement, whereas haemangiomas have a late but very dense appearance due to their pooling of blood.

Scanning the liver in the arterial-dominated phase requires a measurement of the bolus arrival time in the abdominal aorta near the coeliac artery. As in the heart protocols, high flow rates of contrast medium are needed to achieve a compact bolus of contrast. Nevertheless, at least the caudal images of the examination will also have opacification of the portal vein, which is filled rapidly via the splenic vein.

To scan in a predominantly portal phase with contrast application via an antecubital vein, a delay of at least 70 seconds before the scanning begins is recommended. A measurement of portovenous circulation time is not feasible with the methods discussed above, so the timing becomes less exact. At least 140 ml of contrast medium should be applied for this delayed scanning.

The kidneys and the retroperitoneum

The main reason for scanning the kidneys is to exclude a renal cell carcinoma (RCC). This tumour displays a huge amount of vascularization and arteriovenous shunts in the majority of cases. This fact may help to differentiate an RCC from other tumours of the kidney. The scanning should be done in the specific field of view to show only the kidneys.

After measurement of the bolus arrival time in the region of the renal arteries, a high flow bolus of 4 to 6 ml/s is applied. The kidneys are scanned in the timed single-slice mode, using the same milliamperage as in scanning of the liver. Contrast medium injection is terminated when the examination is finished.

An RCC usually shows a hyperintense enhancement compared with the parenchyma of the kidney in the early phase of scanning.

Head and neck; pelvis

For these regions of the body, contrast media protocols can be the same as in conventional spiral CT scanning. Due to the faster scanning times, the contrast flow may be higher and the amount of contrast medium lower.

CONCLUSION

Contrast protocols in EBCT differ from contrast protocols in spiral or conventional CT scanning. The use of high flow rates, high iodine content and monophasic injection protocols are basic requirements for EBCT scanning. In the special, fast-scanning multi-slice mode, the measurement of the bolus arrival time is mandatory. In our opinion, EBCT, with individual timing of the bolus through measurement of the circulation time, can reduce contrast medium consumption and hence dose-dependent side-effects of contrast media. Similarly, EBCT can also help to reduce the costs of CT examinations.

REFERENCES

1. Stanford W, Rooholamini SA, Galvin JR. Ultrafast computed tomography in the diagnosis of aortic aneurysms and dissections. J Thorac Imaging. 1990; 5: 32–39.
2. Thompson BH, Stanford W. Utility of ultrafast computed tomography in the detection of thoracic aortic aneurysms and dissections. Semin Ultrasound CT MRI. 1993; 2: 117–128.

P. Dawson and W. Clauss, (eds.), Advances in X-Ray Contrast: Collected Papers. 74–75
© *1998 Kluwer Academic Publishers.*

Electron beam computed tomography (EBCT)

This review of EBCT is in five parts. Parts 3, 4 and 5 were first published in Advances in X-Ray Contrast. 1995;3:10–11, 12–14 and 15–16.

3. Diagnosis of aortic aneurysms by EBCT

R Knapp, I Bangerl, D zur Nedden
Universitätsklinik für Radiodiagnostik, Radiologie II, Anichstrasse 35, A-6020, Innsbruck, Austria

INTRODUCTION

The incidence of aortic aneurysms in the western population is estimated to be between 2 and 4% [1]. Almost 50% of thoracic aneurysms rupture, a catastrophic event for the patient. The major cause of aortic aneurysms is atherosclerosis, followed by trauma, Marfan's syndrome, and infection (nowadays mostly mycotic) occurring in patients during immunosuppressive therapy. Whereas traumatic aneurysms are almost always false aneurysms, or dissections, mycotic aneurysms or the aneurysms of Marfan's syndrome involve all layers of the vessel and are always true aneurysms. The atherosclerotic type of aneurysm may be a true or a dissecting aneurysm. Dissecting aortic aneurysms are classified in terms of site of origin, after De Bakey:

- Type 1: The dissection begins at the aortic root and reaches down to the descending aorta. The supra-aortic vessels may themselves show a dissection.
- Type 2: Similar to Type 1, but the dissection stops at the brachiocephalic artery.
- Type 3: The dissection begins distal to the left subclavian artery (Figure 1).

This classification is important, both for the surgeon when planning an operation, and also because the three types of dissection can be identified by different diagnostic techniques.

Due to the motion of the aorta, the proximal part of the vessel involved in Type 2 dissection is better

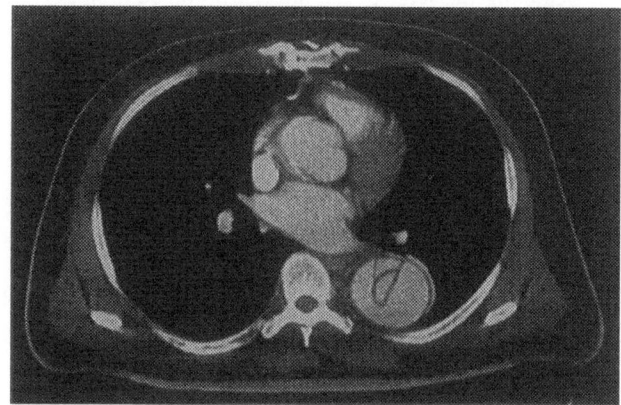

Figure 1   EBCT scan of a Type 3 aortic dissection

visualized by angiography than by CT, whereas the distal parts of the aorta involved in Type 2 and Type 3 dissections are better depicted by CT than by angiography. This is true if conventional CT scanners are used, where the scan time per slice (>1 second) causes a lot of image blurring, due to motion artifacts. As a result, intimal flaps are usually missed in the proximal aorta [2].

Electron beam computed tomography (EBCT), with its superior scanning speed is the method of choice for the non-invasive diagnosis of a Type 1 dissection. The EBCT scanner can operate at a scan time of 100 milliseconds, and therefore it makes sense to trigger the machine by the R-wave of the ECG.

In a series of 17 patients with surgically proven Type 2 dissections, reported by Hamada in 1992, all intimal flaps could be detected correctly by EBCT [2].

SCANNING PROTOCOLS

1. Single-slice mode

The rapid rate of data acquisition in EBCT scanning is optimally obtained by using an ECG trigger pulse to start the scanner at 80% of the R-R interval, which includes end-diastole. Although the 100 millisecond single-slice mode uses only 68 milliamperes per slice and may cause some noisy images, the superior scanning speed and its temporal resolution are mandatory for depiction of the intimal flap. The examination is begun at the level of the intrathoracic supra-aortic vessels, where dissection may be already present. Slice thickness is limited to 6 mm with table increments of 6 mm.

The scanning should be continued along the abdominal aorta until the level of the bifurcation is reached. Apnoea is not necessary – the scanning can be done during shallow breathing without any image degradation caused by artifacts due to respiration.

Contrast protocol

For this scanning, protocol total opacification of the aorta is mandatory. This is guaranteed if 1 ml/s of contrast medium is given for 60 seconds before the scanning starts. Such a long delay has the advantage of providing the liver and the kidneys with a good degree of contrast. However, this does not occur in situations where one of the feeding arteries to these organs is involved in the dissection when normal opacification will not occur.

If the distinction between the false and the true lumen is of interest, an additional scanning protocol in the flow mode can be added.

2. Flow mode

During this procedure, the scanner operates in the multislice mode, where more than one target ring is swept sequentially. As a consequence, two levels of the aorta can be imaged during one 50 millisecond sweep of the scanner. If this process is repeated every other heartbeat, the passage of a bolus of contrast medium through the aorta can be imaged. Usually, the true lumen of the dissecting aneurysm is enhanced earlier than the false lumen.

Contrast protocol

For this mode, measurement of the circulation time is mandatory. We measure the arrival time of the bolus using a test bolus of 12 ml of contrast medium, given at a flow rate of 4 ml/s. Contrast medium is administered in a 'short' bolus of 80 to 100 ml, given at a rate of 4–6 ml/s. Scanning is started four heartbeats before the end of the bolus arrival time.

CONCLUSION

EBCT offers the best way of diagnosing aortic aneurysms and dissections, even in the proximal aorta. The diagnostic value of the method has to be proven in larger series to be equivalent to that of angiography. In the critically ill patient with aortic rupture, angiography is still the method of choice for preoperative diagnosis [3].

REFERENCES

1. Stanford W, Rooholamini SA, Galvin JR. Ultrafast computed tomography in the diagnosis of aortic aneurysms and dissections. J Thorac Imaging. 1990; 5: 32–39.
2. Hamada S. Type A aortic dissection: evaluation with ultrafast CT. Radiology. 1992; 183: 155–158.
3. Thompson BH, Stanford W. Utility of ultrafast computed tomography in the detection of thoracic aortic aneurysms and dissections. Semin Ultrasound CT MRI. 1993; 2: 117–128.

P. Dawson and W. Clauss, (eds.), Advances in X-Ray Contrast: Collected Papers. 76–78
© *1998 Kluwer Academic Publishers.*

4. Imaging the heart by EBCT

R Knapp, I Bangerl, D zur Nedden
Universitätsklinik für Radiodiagnostik, Radiologie II, Anichstrasse 35, A-6020, Innsbruck, Austria

In addition to the superior speed of scanning which allows the creation of almost real-time images of the heart, the EBCT machine is the only CT scanner that enables patient positioning along the short and long axis of the heart. This is possible because the scanner incorporates a specially designed couch which can be slewed up to 25 degrees sideways and which allows the patient's head to be raised by as much as 17 degrees.

In contrast to the widely accepted and available technique of echocardiography, EBCT presents no problems with uncooperative patients or with poor acoustic windows, caused by emphysema or obesity. Further, the problem of interobserver variability is minimized, due to standardized imaging parameters and scan orientation.

PERICARDIUM

The pericardium and the pericardial cavity can be visualized by EBCT, using the single-slice mode at 100 ms scan time. The images should be obtained using 3 or 6 mm beam collimation and with the ECG trigger set at 80% of the R-R interval, to freeze cardiac motion in end-diastole.

Imaging should be performed with angulation of the couch at 25 degrees to the left and the head position raised to 15 –17 degrees [1]. This guarantees that the long axis of the heart is brought parallel to the scan plane. The imaging protocol should involve imaging before and after the use of intravenous contrast media. The reasons for this are two-fold:

1. Non-contrast medium images enable an assessment of the type of pericardial effusion. Whereas transudates will be isodense with, or slightly denser than, water, they will be of more than 30 Hounsfield units because of their higher protein content. The highest densities will be found with haemorrhagic effusion or bleeding inside the pericardial sac. This is a result of the high iron content of the haemoglobin-containing fluid.

2. Small amounts of pericardial calcification may be missed in the presence of contrast agent. EBCT, as an X-ray-based method, is highly sensitive in displayinging calcium and measuring the amount of calcium present.

On the contrast agent-enhanced images, it is easy to define borderlines between the pericardium, the pericardial fluid and the myocardium. Masses inside the pericardium, such as metastases or tumour invasion, can be distinguished. A thickened pericardium is easily depicted.

Contrast administration should provide good opacification of the ventricular cavities throughout the examination.

MYOCARDIUM

Myocardial mass and shape

An estimation of myocardial mass is made by scanning along the short axis of the heart (couch slewed 25 degrees to the left) in single-slice mode, with a scan time of 100 ms. ECG-triggering is mandatory at 80% of the R-R interval. Contrast administration should be done in the same way as for imaging of the pericardium. After the whole heart has been scanned in this manner, it is possible to calculate the myocardial mass for each slice interactively by means of the computer software of the scanner. This is done by tracing the borders of the myocardium using a cursor [2]. The calculation of the overall myocardial mass is done according to a modified Simpson's rule (stack of coins) [1].

In addition to the possibility of measuring the overall myocardial mass, it is also feasible to visualize localized changes of the myocardial wall, such as septal defects or localized thickening, e.g. in hypertrophic obstructive cardiomyopathy. In particular, scars or the formation of an aneurysm after an infarct can be detected easily. The rare entity of myocardial dysplasia in the right ventricle can also be detected because of the high fat content in the right myocardial wall [3]. For such qualitative study of the myocardium it is helpful to image the heart in the scan plane which demonstrates the pathology of the myocardium parallel to the scan plane.

Myocardial function

If the images are obtained in a multislice mode and every target creates images with a frequency of 17/s, triggered at 0% of the R-R wave interval, a cineloop is created. The impression of a thinned myocardium seen in the single-slice mode may prove to be scar tissue if dyskinesia or akinesia is present. Paradoxical movement of an aneurysm is clearly visualized. These qualitative images can be quantified by means of a special algorithm which measures segmental myocardial contraction at every level of the examination (Figure 1).

Scan orientation should be calculated according to location of the infarct. Apical infarcts or aneurysms are best seen in the long-axis view, whereas anteroseptal or true posterior infarcts are best seen when the heart is scanned along the short axis. By measuring the volume of the ventricular cavities in end-diastole and end-systole, stroke volume and ejection fraction can be measured [4]. The use of intravenous contrast media should guarantee good opacification of the ventricular chambers and atria, to depict the borderline between the cavity and the myocardial wall.

Myocardial perfusion

Due to the high spatial and contrast resolution and the absence of motion artifacts, EBCT has the potential to estimate myocardial perfusion. The tracer used for this technique is usually an iodine-containing, non-ionic contrast medium. In contrast to the tracers used in nuclear medicine, such contrast media have no intracellular uptake. As a consequence, the rationale for the estimation of myocardial perfusion by EBCT is based on the distribution of contrast agent in the

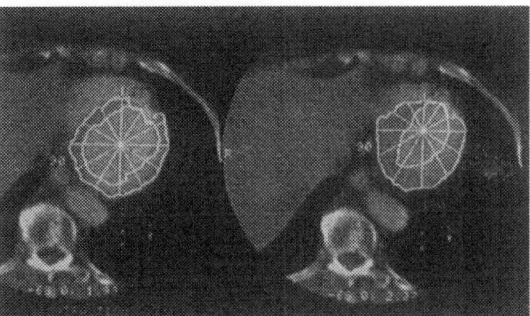

Figure 1 Displaying the movement of the myocardial wall in every segment throughout a cardiac cycle

extracellular compartments of the myocardial wall. The myocardial wall has to be seen as a three-compartment model [5–7], as follows:

1. the vascular bed of the myocardium

2. the interstitial space of the myocardium

3. the muscle cells of the myocardium.

The vascular bed of the myocardium shows early enhancement, whereas the interstitium shows later enhancement. The cellular pool of the myocardium shows no enhancement. The density measured in the myocardium is a summation effect of the fractional distribution of the three zones in the myocardial wall. Therefore, early and strong enhancement of the image is a good indicator of a normal myocardium. An ischaemic myocardium shows a weaker image enhancement in the early phase, due to less vascularization of the tissue. In the late phase, enhancement can reach the same levels as in the normal myocardium because the reduction of the vascular bed is compensated for by enhanced interstitial space. An infarcted myocardium may show early enhancement, comparable to the levels of an ischaemic myocardium. The late phase of infarcted myocardium may show higher levels of image enhancement. This can be explained as a loss of both the vascular bed and the cellular compartment of the myocardium. Both compartments have been replaced by the 'interstitium' exhibiting later enhancement, caused by diffusion of contrast agent. As absolute CT numbers are of little value in estimating the extent of

78

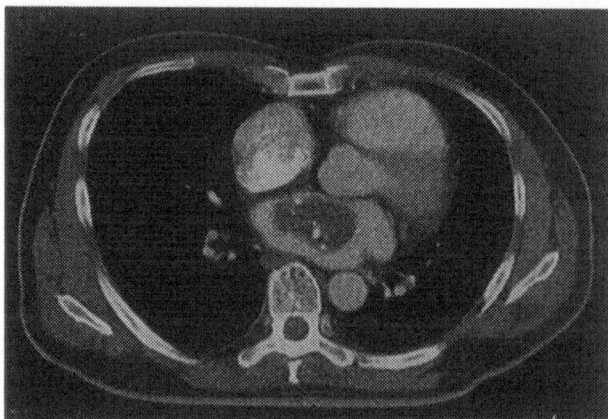

Figure 2 Large atrial mass (myxoma)

myocardial perfusion, the best parameter for categorizing the myocardium, as explained above, is the calculation of the ratio of myocardial CT number to luminal CT number of the left ventricle.

ENDOCARDIUM

Even though the endocardium and the valves cannot be visualized constantly, it is possible to estimate the degree of any regurgitation by a comparison of the stroke volumes of both ventricles, which should be almost identical in a healthy individual. Any difference between the stroke volumes allows an estimation of the regurgitation volume [7].

Endocardial thrombi are often seen adjacent to scar tissue after myocardial infarction and may be differentiated from endocardial masses, such as myxomas, because of their uptake of contrast agent (Figure 2).

Since myxomas are often located in the left atrium, they are poorly visualized by means of transthoracic echocardiography, and special techniques, such as transoesophageal ultrasound, are required.

CONTRAST MEDIA APPLICATION

The use of contrast media in EBCT for the heart is similar to that for pulmonary embolism and an explanation of this can be found in the following Section 5.

CONCLUSION

EBCT provides a good method for the non-invasive imaging of the heart. Results are comparable to those of standard techniques and are reproducible. The method does not suffer from poor image quality in obese patients or in patients with emphysema. The short acquisition time for the whole dataset makes it possible to examine patients who, because of their poor clinical condition, are unable to cooperate for a longer time.

REFERENCES

1. Rumberger JA, Sheedy PF, Stanson WA. Ultrafast (Cine) CT:Advantages and limitations in cardiovascular assessment. Internal Medicine. 1988; 6: 2–7.
2. Rumberger JA, Lipton MJ. Ultrafast cardiac CT scanning. Cardiol Clin. 1989; 8: 713–734.
3. Hamada S, Takamiya M, Ohe T. Arrythmogenic right ventricular dysplasia: Evaluation with electron-beam CT. Radiology. 1993; 6: 723–727.
4. Marcus LM, Rumberger JA, Stark CA, et al. Cardiac applications of ultrafast computed tomography. Am J Cardiac Imaging. 1988; 6: 116–121.
5. Naito H, Saito H, Takamiya M, et al. Quantitative assessment of myocardial enhancement with iodinated contrast medium in patients with ischemic heart disease by ultrafast X-ray computed tomography. Invest Radiol. 1992; 6: 436–441.
6. Ludman PF, Coats AJ, Burger P, et al. Validation of measurement of regional myocardial perfusion in humans by ultrafast X-ray computed tomography. Am J Cardiac Imaging. 1993; 12: 267–279.
7. Marcus ML, Stanford W, Hajduczok ZD, et al. Ultrafast computed tomography in the diagnosis of cardiac disease. Am J Cardiol. 1989; 9: 54E–59E.

P. Dawson and W. Clauss, (eds.), Advances in X-Ray Contrast: Collected Papers. 79–80
© 1998 Kluwer Academic Publishers.

5. Diagnosis of pulmonary embolism by EBCT

R Knapp, I Bangerl, D zur Nedden
Universitätsklinik für Radiodiagnostik, Radiologie II, Anichstrasse 35, A-6020, Innsbruck, Austria

INTRODUCTION

Because of the lack of specificity of associated clinical symptoms, pulmonary embolism (PE), a serious event, needs a reliable tool for its diagnosis. If PE is not diagnosed and treated correctly, the patient may suffer relapse after the first episode which may result in death; an adequate diagnosis is therefore mandatory. An accurate diagnosis of PE is also needed because of the relative risk of the treatment – anticoagulant therapy has an estimated complication rate of up to 15 % [1].

The diagnostic technique should be easy to perform, low cost and readily available, because of the frequency of clinically suspected PE. The best method until the early 1990s was the ventilation/perfusion scan. Although cheap and widely available, the method lacks specificity and fulfils only the criteria for an exclusion diagnosis of PE. The gold standard in the diagnosis of PE is still pulmonary angiography. The disadvantages of angiography are its invasiveness and a complication rate of appproximately 6.5%, together with its lack of availability [2]. Another problem is the difficulty in interpretation of pulmonary angiograms due to a high degree of superposition and breathing artifacts. The value of spiral CT in the diagnosis of PE was highlighted in a report by Remy-Jardin in 1991 [3]; the sensitivity and specificity of this single breathhold technique were stated to be 100% and 96%, respectively. The problem with volumetric CT is that the single breathhold technique is often impossible with non-co-operative patients or patients unable to hold their breath. The best conditions for scanning the pulmonary arteries are provided by EBCT. The EBCT scanner is capable of scanning at a speed of 100 milliseconds per slice, which enables it to operate in less than the time of one R-R interval. An ECG-triggering of the individual slice at end-diastole, achieved at 80% of the R-R interval, generates data without any interference from motion artifacts and the pulmonary arteries can thereby be viewed without any blurring due to cardiac and vessel movement.

EBCT PROTOCOL FOR PE

For the diagnosis of PE, the entire thorax should be examined. A restriction of the field of view as proposed by Teigen et al. is not desirable in our opinion [3]. Slice thickness is 6 mm and the table increment is also 6 mm. Scanning is triggered at 80% of the R-R interval, and patients are recommended to take shallow breaths throughout the examination.

CONTRAST MEDIA APPLICATION

If the patient's clinical condition is not serious, a measurement is made of the circulation time using a test bolus of 12 ml of non-ionic contrast agent with an iodine content of 370 mg/ml. The test bolus is administered via an antecubital vein at a flow rate of 4 ml/s. The bolus arrival time is measured in the right ventricle by the computer software of the scanner and can be displayed as a time-density curve. For the examination, 2–4 ml/s of contrast medium are applied, depending on the bolus arrival time and the size of the patient. If the patient has a short circulation time, a higher flow rate is used, to guarantee a minimum amount of intravascular contrast before the scanning starts. Scanning is started with a delay of bolus arrival time +10 s. The contrast medium injection is terminated when the number of scans remaining is equivalent to the bolus arrival time in seconds. If the patient's condition is not good enough, scanning is started immediately without measuring the bolus arrival time. The bolus arrival time at the right heart is estimated to be 15 s in these cases. If the patient's clinical condition indicates right heart failure, 10 s are added for the symptom of neck-vein congestion, another 10 s for tachypnoea or dyspnoea, and a delay

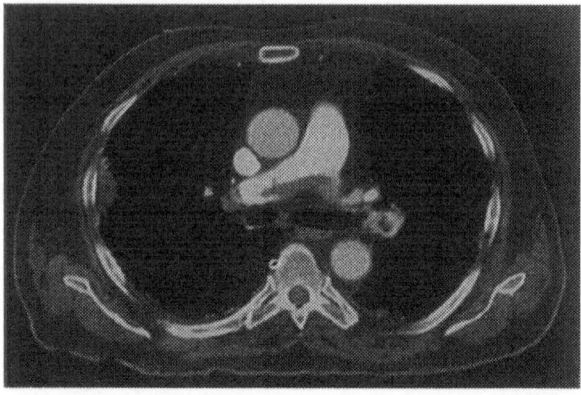

Figure 1 Pulmonary embolism in both pulmonary arteries

of 60 s in the unconscious patient. A flow rate of 2 ml/s is used in most cases, less than 100 ml of contrast medium is required to obtain good results.

IMAGE INTERPRETATION

The acquired data should be displayed with a narrow window setting usually used for imaging the mediastinum. In these images, filling defects inside the pulmonary arteries can be seen up to the subsegmental pulmonary arteries (Figure 1).

Acute PE typically occludes one or more pulmonary vessels in the periphery of the pulmonary vasculature. In chronic PE, the embolus is incorporated in the wall of the vessel, and contrast medium by-passes it to the distal pulmonary arteries. For a more reliable diagnosis, all images should also be viewed using the lung window technique. A zone of consolidation in the periphery of the lung, without an airbronchogram and/ or with a pleural effusion, can be indicative of a very small embolus in the periphery of the lung. In doubtful cases, a pulmonary angiogram should be performed.

CONCLUSION

EBCT, with an appropriate scanning protocol, provides the radiologist with optimal diagnostic information about PE. This has been proven by Teigen et al. in 60 patients who underwent EBCT and pulmonary angiography as the gold standard in the diagnosis of PE [1]. In their series, the diagnostic sensitivity and specificity reached almost 100%. The role of the ventilation-perfusion scan is primarily to exclude PE if the scan is negative.

REFERENCES

1. Teigen CL, Maus TP, Sheedy PF, et al. Pulmonary embolism: diagnosis with contrast-enhanced electron-beam CT and comparison with pulmonary angiography. Radiology. 1995; 194: 313–319.
2. Stein PD, Athanasoulis C, Alavi A, et al. Complications and validity of pulmonary angiography in acute pulmonary embolism. Circulation. 1992; 85: 462–468.
3. Remy-Jardin M, Remy J, Wattine L, et al. Central pulmonary thromboembolism: diagnosis with spiral volumetric CT with the single-breath-hold technique – comparison with pulmonary angiography. Radiology. 1992; 185: 381–387.

P. Dawson and W. Clauss, (eds.), Advances in X-Ray Contrast: Collected Papers. 81–88
© 1998 Kluwer Academic Publishers.

Iodinated contrast agents in neuroradiology

Ronald J. Zagoria, MD

Department of Radiology, Bowman Gray School of Medicine, Medical Center Boulevard, Winston-Salem, North Carolina, 27157-1088, USA

INTRODUCTION

Iodinated contrast materials, which are commonly used in neuroradiological applications, are an unusual and somewhat enigmatic class of drugs. We frequently give intravascular injections of these agents in large volumes, yet they have no therapeutic value. Their value lies instead in their ability to attenuate X-rays, a property imparted to them only by the constituent iodine atoms that make up a small fraction of their molecules. The ideal model contrast medium would yield superb radiographic contrast, would be thoroughly soluble and have low viscosity, would be rapidly excreted from the body, and would have no toxic side effects. Although currently available iodinated contrast agents approach these ideals, some limitations persist.

HISTORY

The first radiographic contrast agents were gases, including room air, oxygen and nitrous oxide, used experimentally by Dandy and Jacobeus in 1919. These agents were unsuitable for intravascular applications. During the 1920s, Moses Swick developed the first iodinated water-soluble contrast material for intravascular injection in humans [1,2]. This agent was based on a pyridone molecule with a single iodine substituent and was marketed as 'Uroselectan'. It, and similar agents with slight modifications, was in use until the mid-1950s, when newer iodinated contrast materials based on a substituted benzene ring were developed. Two substituted benzoic acids came into common usage: diatrizoic acid and iothalamic acid. These fully substituted benzoic acids incorporate three iodine molecules per unit and are completely

dissociated in solution so that they are water-soluble. The cations used commonly with ionic contrast materials to form salts are sodium and meglumine, which can be used alone or in combination, depending on the formulation of the manufacturer [3]. These ionic agents have osmolalities five to 10 times that of normal serum [3]. Osmolality is a measure of the number of particles dissolved in the solution. Osmotic pressure, a major source of adverse reactions to radiographic contrast materials, is determined by the number of particles present per unit volume of solution. Osmotic pressure controls the passive distribution of water between intracellular and extracellular spaces. One contribution to hyperosmolality in ionic contrast materials is from the cations in solution, which account for 50% of the overall osmolality of the contrast material. Unfortunately, the cations yield no diagnostic information, as they do not effectively attenuate X-rays.

Realization by several researchers that marked hyperosmolality of contrast materials was a source of adverse reactions led to the development of low-osmolar agents. The first of these agents, metrizamide, was developed by Torsten Almén [4]. Although this agent was inconvenient and expensive for routine intravascular applications, it was used extensively for myelography because of its favourable neurotoxic profile. During the 1980s, several new non-ionic, low-osmolar contrast materials were developed. These compounds are not salts but are soluble in water as a result of the inclusion of extensive hydrophilic hydroxyl side chains attached to fully substituted benzene rings containing three iodine molecules. Agents in this group include iohexol, iopamidol, iopromide and ioversol. One additional low-osmolar

Part of this article has appeared previously and is reprinted with permission from: Zagoria RJ. Iodinated contrast agents in neuroradiology. In: Elster AE, ed. Contrast Agents in Neuroimaging, Neuroimaging Clinics of North America, WB Saunders Co., 1994: Vol 4(1).

82

agent, an ionic dimer, became available during the same period. This agent, ioxaglate, has slightly lower osmolality than non-ionic agents with comparable iodine concentrations and is useful for intravascular applications that would otherwise be more painful [5]. Unfortunately, ioxaglate has a high chemotoxicity and requires the inclusion of a cation, the addition of which, it has been suggested, may lead to an increase in the rate of some adverse reactions [6].

Enthusiasm for these new low-osmolar, iodinated contrast materials has been great. Their higher biocompatibility and patient tolerance, in comparison with that of ionic agents [7], has led to widespread use of these improved formulations. However, universal use of low-osmolar contrast materials has been delayed as a result of the sizable difference in cost between the relatively expensive low-osmolar agents and the older ionic agents [8,9]. While issues regarding appropriate applications of new low-osmolar agents persist, even newer agents are in various stages of development. Some of these new agents are lower osmolar non-ionic dimers. The incremental margin of safety of these newer agents remains to be determined.

ADVERSE CONTRAST MEDIA REACTIONS

Adverse reactions are initiated by one of two factors, chemotoxic and idiosyncratic phenomena [10]. Chemotoxic reactions relate to the chemical structures of the contrast molecules. Signs and symptoms of idiosyncratic reactions may be similar to those of allergic or anaphylactic reactions. The exact basis for either class of adverse reaction remains unclear.

Chemotoxic effects

Chemotoxic reactions can affect virtually all organ systems. The ionic nature of many contrast materials may transiently alter the extracellular fluid make-up and alter membrane potentials, which can be affected by soluble ions [11]. In addition, some authors believe that the calcium-binding characteristics inherent in ionic contrast materials and enhanced in some by the addition of chelating agents play a major role in the induction of adverse reactions [11]. Sudden reductions in ionic calcium can interfere with the electrical conduction mechanisms of the cardiovascular system

and lead to cardiac dysfunction. In addition, all contrast media interfere to some extent with coagulation and clotting factors [12]; in standard doses, this interference is usually clinically insignificant. It is clear that ionic agents have a more profound effect on inhibiting blood coagulation [13] than non-ionic contrast agents. However, the lessened anticoagulant effect of non-ionic contrast agents does not appear to increase the risk of thromboembolic events associated with angiographic procedures if meticulous catheter and guidewire techniques are employed [14,15]. Adverse chemotoxic side effects also include red blood cell damage, myocardial depression, cardiac arrhythmia, renal toxicity and bronchospasm [10,11].

Neurotoxicity

The central nervous system is well insulated from chemotoxic effects of intravascular contrast material by the presence of the blood-brain barrier, which protects the delicate environment of the central nervous system from potentially detrimental materials. Normally, the blood-brain barrier is impermeable to iodinated contrast material [16], but its integrity can be compromised by damage to the endothelium of cerebral capillaries. This damage can occur solely as a result of the hyperosmolality of the contrast material [16], and damage appears to be lessened with the use of low-osmolar agents [17]. Intravascular contrast material comes into contact with the central nervous system in areas where the blood-brain barrier is incomplete (choroid plexus, stalk of the pituitary, area postrema, pineal body, medial eminence of the neural hypophysis, subfornical and subcommissural organs, and organum vasculosum) [17]. Contact with radiographic contrast materials at these sites is a suggested source for supposed central nervous system-mediated systemic reactions to contrast material. In addition, pathologic nervous system lesions often increase the permeability of the blood-brain barrier [17–19]. Neoplasms in the central nervous system, as well as infections and infarctions, can increase leakage of radiographic contrast material across the blood-brain barrier. Although this leakage has the benefit of increasing the visibility of these pathologic lesions on a contrast-enhanced central nervous system scan, a side effect is increased exposure of normal neuronal tissues to the neurotoxic effects of radiographic contrast materials. Neurotoxicity is related to both concentration and type of ions present in solution. The presence of electrically charged ions can inhibit normal neurotransmission and elicit abnormal neurostimulation, leading to the development of seizures [17]. Non-ionic agents have minimal electrical effects. In addition, diatrizoate anions have been shown to be more neurotoxic than those of similar iothalamate-based ionic agents used in neuroradiology [11]. Contrast media-related neurologic side effects include seizures, cortical blindness, paresis and encephalopathy. The major risk factor for developing a central nervous system chemotoxic side effect is the presence of an underlying lesion of the central nervous system.

The risk factors for other chemotoxic reactions include renal insufficiency, diabetes mellitus, severe cardiovascular or pulmonary disease, and generalized debilitation [20–22].

Idiosyncratic reactions

Idiosyncratic reactions are those that mimic anaphylactic responses. They range from the mild, including hives and mild bronchospasm, to the severe, including sudden death [10,11]. The aetiology of these idiosyncratic reactions is uncertain, but proposed mechanisms include histamine release, complement cascade activation, and direct central nervous system effects [10]. It is clear that they are not purely allergic reactions, as no circulating IgE antibodies to contrast material can be isolated in patients with a history of idiosyncratic contrast material reaction. There are known risk factors for idiosyncratic reactions, including asthma, various allergies, such as hay fever and drug or food allergies, and previous minor reactions to contrast material [23,24]. Patients with any of these risk factors have a five- to ten-fold greater risk than the general population of developing an idiosyncratic reaction to administered intravascular contrast material [25]. For patients who have had a previous severe idiosyncratic reaction to contrast material, the risk of developing another idiosyncratic reaction to administered ionic contrast material is as high as 40%. Administering steroids and using non-ionic contrast material greatly reduces the risk of idiosyncratic reactions in these patients [26]. Prophylactic treatment is discussed later in this article.

ADVERSE REACTION RATES

For both categories of contrast media reaction, the injection of ionic contrast materials is associated with adverse reactions in 5 to 10% of patients [24]. Most of these reactions are mild and include sneezing, coughing, rhinitis, conjunctival oedema, mild urticaria, pruritus, vomiting and lightheadedness. Severe reactions occur in approximately one in 500 injections of ionic contrast material [24], and the associated death rate ranges from one in 40 000 to one in 100 000 injections [23,24]. The overall rate of adverse reactions with non-ionic contrast materials is approximately 1/6 that of ionic agents [23]. The overall rate of severe reactions associated with the use of non-ionic contrast materials is one in 2500 injections [23].

Non-ionic versus ionic contrast materials

It is clear that non-ionic, iodinated contrast materials are better tolerated and have substantially lower adverse reaction rates than similar ionic agents. These data have been substantiated with both animal models and large series examining contrast media reaction rates in patients [10,23,27]. The use of non-ionic contrast materials does not compromise diagnostic efficacy; in fact, the reduction in pain and adverse side effects may lessen patient motion and therefore improve the diagnostic images in some studies. Although it is clear that with non-ionic agents there are fewer adverse reactions, both idiosyncratic and chemotoxic, and that these agents are much less painful for arteriography, they are not universally accepted. Currently, in the United States, non-ionic agents cost 10 to 25 times as much as similar ionic contrast material. This discrepancy in expense has inhibited the universal use of non-ionic contrast materials. Choosing between ionic and non-ionic contrast materials remains difficult. The advantages of the improved safety profile associated with non-ionic agents must be weighed against the cost savings associated with ionic agents.

In neuroimaging, the injection of contrast materials into the carotid circulation induces vasodilatation, which can lead to systemic effects. Stimulation of the carotid body alters heart rate and systemic blood pressure [28]. In addition, intravascular contrast material injected for either cerebral arteriography or cranial computed tomography can lead to central nervous system neurotoxicity if the permeability of the blood-brain barrier is increased. It has been demonstrated that non-ionic agents injected into the carotid artery cause substantially fewer reflex haemodynamic effects than do ionic contrast material injections [28], and direct neurotoxicity is greatly lessened by the use of non-ionic agents [17].

PROPHYLACTIC TREATMENT

Measures to prevent adverse reactions to contrast media should be undertaken whenever possible. One such measure is the use of non-ionic contrast materials in high-risk patients, if not employed universally. Another is prophylactic treatment for patients who have experienced previous major adverse reactions to contrast materials. Studies have demonstrated that up to 40% of these patients will experience a subsequent adverse reaction when exposed again to ionic contrast material. Recent clinical studies have claimed to show that the use of preprocedure steroids in combination with antihistamines and non-ionic contrast materials can prevent virtually all significant adverse reactions in this group of patients [25,26]. However, the administration of steroids must precede the injection of contrast material by a number of hours to be effective. The current steroid regimen recommended for patients who have had a previous major reaction to contrast material includes the administration of 50-mg prednisone tablets at 6-hour intervals, beginning 13 hours prior to the scheduled contrast material injection [26]. This approach will allow a minimum of three doses of steroids to be ingested prophylactically. One 50-mg dose of diphenhydramine should also be administered 30 to 60 minutes prior to intravascular contrast material administration [26]. Finally, only non-ionic contrast material should be employed in these high-risk patients [25,26]. With this protocol, the rate of adverse reactions should be well below 1%, and virtually all life-threatening reactions should be avoided [26].

NEPHROTOXICITY

Iodinated contrast materials are nephrotoxic. The development of clinically evident contrast-material nephrotoxicity probably results from a number of factors. The effect of intravascular contrast material on the renal vascular supply is biphasic. Initially, there is a mild vasodilatation, which is followed rapidly by a more prolonged vasoconstriction. In some patients, this vasoconstriction may lead to renal ischaemia and breakdown of renal basement membrane junctions [29–31]. As a result, the tubular epithelium is exposed to the direct toxic effects of contrast material already within the renal tubules, further increasing the damaging effects of the contrast material.

Recent studies have clarified the rate and risk factors for developing contrast media nephrotoxicity. Risk factors include renal insufficiency, diabetes mellitus and, possibly, congestive heart failure [32,33]. Patients without any of these risk factors experienced some renal nephrotoxicity in 2 to 5% of exposures [32,33]. Patients with one or more risk factors had a 9 to 16% rate of substantial nephrotoxicity [32,33]. None of these patients required dialysis, and there was no detectable difference in clinically significant contrast-medium nephrotoxicity between ionic and non-ionic contrast materials [32,33]. Other studies have suggested that the use of non-ionic agents does lead to a reduction in nephrotoxicity at a subclinical level [34]. The significance of this finding remains questionable. The use of moderate doses of contrast material appears to be safe in patients with serum creatinine levels of 1.5 mg/dl or less. The indications for contrast material injections in patients with elevated serum creatinine levels should be evaluated critically, and other diagnostic options should be considered. If contrast material administration is deemed necessary, then the patient should be hydrated both before and after the procedure, to minimize side effects.

TREATMENT OF ADVERSE REACTIONS

A thorough description of current recommendations for the treatment of adverse contrast media reactions is beyond the scope of this article. Several recent articles have outlined rational approaches for management of such situations [9,35,36]. Suffice it to say that radiologists employing iodinated contrast materials for intravascular use must be prepared to treat adverse reactions. Ready availability of monitoring devices, resuscitation equipment and commonly used medications is mandatory [9]. The most common severe anaphylactoid reactions associated with intravascular contrast media administration include tachycardia, hypotension, and some element of bronchospasm. In these cases, early recognition is extremely important. Once a reaction is recognized, therapy must be administered to reverse it. The rapid infusion of intravenous fluids such as normal saline or lactated Ringer's solution is mandatory and should be augmented with facial administration of oxygen and the injection of epinephrine and antihistamines. If bronchospasm is the most prominent feature or is refractory to other treatment, a β_2-agonist inhaler is often effective in reversing this symptom. Vasovagal reactions are best treated by administering intravenous fluids, placing the patient in the Trendelenburg position, and injecting atropine as needed.

Extravasation of contrast materials into the subcutaneous tissues around the injection site can be problematic. Small volumes of ionic or non-ionic contrast materials are unlikely to cause significant tissue damage. However, larger volumes of ionic contrast material can lead to tissue damage and resulting necrosis [37]. Tissue necrosis of this type has not been reported with the use of non-ionic contrast material, and experimental data suggest that it is relatively innocuous when injected subcutaneously [38]. When contrast material extravasation occurs, the affected limb should be elevated and cold compresses should be applied for a short time period.

SPECIAL CIRCUMSTANCES

Some patients, namely, those harbouring a phaeochromocytoma, those with multiple myeloma, and those with hyperuricaemia, thyroid disease or sickle cell anaemia, necessitate special consideration. At times, it has been suggested that all of these patient groups are at increased risk for developing adverse reactions to contrast media. Specifically, it has been suggested that in patients with multiple myeloma and those with hyperuricaemia, the risk of developing irreversible renal failure is greatly increased. However, further scrutiny of the data regarding hyperuricaemic patients has revealed that these patients are also likely to have coexistent renal insufficiency or another risk factor [29]. Evaluation of previous case reports demonstrates that renal insufficiency alone, rather than

hyperuricaemia, is the independent risk factor for developing contrast nephrotoxicity [29]. Furthermore, currently used contrast materials, excluding cholangiographic media, do not promote hyperuricosuria, the proposed initiator of urate nephropathy [29,39]. Therefore, hyperuricaemia in isolation should not be considered a risk factor for contrast material-induced nephrotoxicity.

There has been speculation that in patients with multiple myeloma, obstructive uropathy can result from intratubular precipitation of abnormal urinary proteins induced by the intravascular injection of contrast material. However, recent reviews have failed to support this conclusion, and it is believed that the risk of contrast-induced nephropathy in patients with multiple myeloma is low [40–42]. A review of several series evaluating this risk demonstrated a 0.6 to 1.25% risk of developing acute renal failure, whereas the risk was 0.15% in a control population [42]. Previous authors have concluded that, although the incidence of contrast medium-induced renal failure in myeloma patients is low, it is probably greater when they are dehydrated [41]. However, contrast media can be administered safely to well-hydrated patients with multiple myeloma [41].

There has also been speculation that contrast material can induce sickle cell crisis in patients with sickle cell disease. Hyperosmolar solutions can induce sickling in at-risk patients [43]. Some authors have suggested that red blood cell crenation caused by intravascular contrast material can stimulate a sickle cell crisis, which can include pain, haemolysis and focal infarction [12,43,44]. Both patients with sickle cell disease (haemoglobin SS) and those with haemoglobin SC disease appear to be at risk [12,43]. Non-ionic contrast media induce much less sickling *in vitro* than do comparable ionic agents [45]. Current information suggests that patients with haemoglobin SS or haemoglobin SC should be well hydrated prior to receiving intravascular contrast material and that only non-ionic contrast material should be employed [45].

Evidence to suggest that iodinated contrast materials can induce acute thyrotoxicosis, both in patients with known thyroid disease and in those with no previous thyroid disease, has been anecodotal [46–48]. However, it is clear that iodinated contrast agents contain – as a result of this production – a very small amount of iodide and undergo some deiodination [46]. In the United States, thyrotoxicosis is a rare complication of contrast media administration, and the resulting hyperthyroidism is self-limiting and should therefore be treated conservatively [46].

Finally, intravascular administration of iodinated contrast material in patients with phaeochromocytomas should be avoided. The induction of a fatal hypertensive crisis is an anecdotal complication of adrenal arteriography and venography in these patients [49,50]. Patients with phaeochromocytomas seem to be at risk, albeit small, of developing a hypertensive crisis induced by the intravenous injection of contrast material for CT scanning [50]. Therefore, if the use of contrast material is mandatory in a patient with a documented phaeochromocytoma, an α-adrenergic blockade should be administered prior to the injection of the contrast material, to guard against a hypertensive crisis.

IODINATED CONTRAST MATERIALS IN PAEDIATRIC PATIENTS

Iodinated contrast agents for intra-arterial and intravenous use in paediatric patients are identical to those employed for radiological procedures in adults. In very young children, non-ionic contrast materials are generally preferred because of their lower osmolality, which lessens the dramatic fluid shifts that may occur. In addition, the decreased pain associated with low-osmolar agents may improve patient compliance more

profoundly in this group than in adult patients. The dose of contrast material used in children depends on the patient's size and the procedure being performed. In general, a total contrast material dose of 4 ml/kg of body weight of medium-to-high density (250 mg I/ml – 370 mg I/ml) contrast material should not be exceeded [51].

CONTRAST MATERIAL DOSES IN NEURORADIOLOGY

There is a wide range of acceptable doses, concentrations and injection methods for contrast media use in neuroradiology. The use of appropriate doses of either low- or high-osmolar agents results in acceptable contrast enhancement. Since low-osmolar agents cause less intravascular dilution, for a given iodine concentration, low-osmolar agents yield greater enhancement for brain CT studies than the same dose of a high osmolar ionic agent. This allows reduced doses of contrast material when using low-osmolar agents [52,53]. Likewise, neuroangiography contrast material doses are variable, depending on personal preference and injection technique. Digital subtraction arteriography requires smaller injected doses of iodinated contrast material than for film-screen technology studies of similar diagnostic quality.

Listed in Table 1 is a summary of the contrast material protocol used at this institution for neuroradiology, including digital neuroangiography.

Table 1   Intravascular contrast material used at Bowman Gray Medical Center

Procedure	Contrast agent	Injection volume	Injection technique
Brain CT	250 mgI/ml	100 ml	IV drip infusion
Neck CT	300 mgI/ml	100–150 ml	1.0 ml/s x 30 s then 0.5 ml/s for 120 ml or 1.5 ml/s x 20 s then 0.5 ml/s for 70 ml (helical CT)
Common carotid arteriography	300 mgI/ml	8–10 ml	6–8 ml/s rate
Internal carotid arteriography	300 mgI/ml	7–8 ml	5-6 ml/s rate
External carotid arteriography	300 mgI/ml	6 ml	3 ml/s rate
Vertebral arteriography	300 mgI/ml	7–8 ml	5 ml/s or 10 ml/s for crossfilling of contralateral vertebro-basilar artery
Aortic arch arteriography	300 mgI/ml	30 ml	15 ml/s rate

REFERENCES

1. Grainger RG. Intravascular contrast media – the past, the present and the future. Br J Radiol. 1982; 55: 1.

2. Swick M. Radiographic media in urology. The discovery of excretion urography: Historical and developmental aspects of the organically bound urographic media and their role in the varied diagnostic angiographic areas. Surg Clin North Am. 1978; 58: 977.

3. Fischer HW. Catalog of intravascular contrast media. Radiology. 1986; 159: 561.

4. Bettmann M. Ionic versus non-ionic contrast agents and their effects on blood components. Invest Radiol. 1988; 23 (Suppl 2): S309.

5. Smith DC, Yahiku PY, Maloney MD, et al. Three new low-osmolality contrast agents: A comparative study of patient discomfort. AJNR. 1988; 9: 137.

6. Nakstad PH, Bakke SJ, Kjartansson O, et al. Omnipaque vs. Hexabrix in intravenous DSA of the carotid arteries: Randomized double-blind crossover study. AJNR. 1986; 7: 303.

7. Dawson P. Iodinated intravascular contrast agents past and present: Toxicity considerations. Invest Radiol. 1990; 25 (Suppl 1): S11.

8. Steinberg EP, Anderson GF, Powe NR, et al. Use of low-osmolality contrast media in a price-sensitive environment. AJR. 1988; 151: 271.

9. Van Sonnenberg E, Neff CC, Pfister RC. Life-threatening hypotensive reactions to contrast media administration: Comparison of pharmacologic and fluid therapy. Radiology. 1987; 162: 15.

10. Dawson P. Iodinated intravascular contrast agents. A review. J Intervent Radiol. 1987; 2: 51.

11. Swanson D. Conventional or low-osmolality: Picking the right contrast media. Diagnostic Imaging. 1988; 10: 191.

12. Rao AK, Thompson R, Durlacher L, et al. Angiographic contrast agent-induced acute hemolysis in a patient with haemoglobin SC disease. Arch Intern Med. 1985; 145: 759.

13. Bettmann M. Clinical summary and conclusions: Ionic versus non-ionic contrast agents and their effects on blood components. Invest Radiol. 1988; 23 (Suppl 2): S378.

14. Dawson P. Contrast agents, red cells, coagulation, and the angiographer. Invest Radiol. 1990; 25 (Suppl 1): S117.

15. Gertz EW. Thromboembolic events and non-ionic contrast. Diagnostic Imaging. 1989: 11: 106.

16. Harnish PP, Hagberg DJ. Contrast media-induced blood-brain barrier damage: Potentiation by hypertension. Invest Radiol. 1988; 23: 463.

17. Stolberg HO, McClennan BL. Ionic versus non-ionic contrast use. Curr Probl Diagn Radiol. 1991; 20: 47.

18. Avrahami E, Weiss-Peretz J, Cohn DF. Focal epileptic activity following intravenous contrast material injection in patients with metastatic brain disease. J Neurol Neurosurg Psychiatry. 1987; 50: 221.

19. Haslam RHA, Cochrane DD, Amundson GM, et al. Neurotoxic complications of contrast computed tomography in children. J Pediatr. 1987; 111: 837.

20. Barrett BJ, Parfrey PS, Vavasour HM, et al. A comparison of non-ionic, low-osmolality radiocontrast agents with ionic, high-osmolality agents during cardiac catheterization. N Engl J Med. 1992; 326: 431.

21. Moore RD, Steinberg EP, Powe NR, et al. Frequency and determinants of adverse reactions induced by high-osmolality contrast media. Radiology. 1989; 170: 727.

22. Steinberg EP, Moore RD, Powe NR, et al. Safety and cost effectiveness of high-osmolality as compared with low-osmolality contrast material in patients undergoing cardiac angiography. N Engl J Med. 1992; 326: 425.

23. Katayama H, Yamaguchi K, Kozuka T, et al. Adverse reactions to ionic and non-ionic contrast media: A report from the Japanese Committee on the Safety of Contrast Media. Radiology. 1990; 175: 621.

24. Shehadi WH. Adverse reactions to intravascularly administered contrast media: A comprehensive study based on a prospective survey. AJR. 1975; 124: 145.

25. Siegle RL, Halvorsen RA, Dillon J, et al. The use of iohexol in patients with previous reactions to ionic contrast material: A multicenter clinical trial. Invest Radiol. 1991; 26: 411.

26. Greenberger PA, Patterson R. The prevention of immediate generalized reactions to radiocontrast media in high-risk patients. J Allergy Clin Immunol. 1991; 87: 867.

27. Palmer FJ. The RACR survey of intravenous contrast media reactions: Final report. Australas Radiol. 1988; 32: 426.

28. Yamashita K, Hayakawa K, Tanaka M, et al. Cardiovascular responses following the intracarotid injections of ionic and non-ionic contrast media compared with various mannitol solutions: Correlation with osmolality. Invest Radiol. 1988; 23: 680.

29. Byrd L, Sherman RL. Radiocontrast-induced acute renal failure: A clinical and pathophysiologic review. Medicine. 1979; 58: 270.

30. Spinler SA, Goldfarb S. Nephrotoxicity of contrast media following cardiac angiography: Pathogenesis, clinical course, and preventive measures, including the role of low-osmolality contrast media. Ann Pharmacother. 1992; 26: 56.

31. Weisberg LS, Kurnik PB, Kurnik BRC. Radiocontrast-induced nephropathy in humans: Role of renal vasoconstriction. Kidney Int. 1992; 41: 1408.

32. Parfrey PS, Griffiths SM, Barrett BJ, et al. Contrast material-induced renal failure in patients with diabetes mellitus, renal insufficiency, or both: A prospective controlled study. N Engl J Med. 1989; 320: 143.

33. Schwab SJ, Hlatky MA, Pieper KS, et al. Contrast nephrotoxicity: A randomized controlled trial of a non-ionic and an ionic radiographic contrast agent. N Engl J Med. 1989; 320: 149.

34. Gomes AS, Lois JF, Baker JD, et al. Acute renal dysfunction in high-risk patients after angiography: Comparison of ionic and non-ionic contrast media. Radiology. 1989; 170: 65.

35. Bush WH, Swanson DP. Acute reactions to intravascular contrast media: Types, risk factors, recognition, and specific treatment. AJR. 1991; 157: 1153.

36. Cohan RH, Dunnick NR, Bashore TM. Treatment of reactions to radiographic contrast material. AJR. 1988; 151: 263.

88

37. Loth TS, Jones DEC. Extravasations of radiographic contrast material in the upper extremity. J Hand Surg [AM]. 1988; 13: 395.

38. Kim SH, Park JH, Kim YI, et al. Experimental tissue damage after subcutaneous injection of water soluble contrast media. Invest Radiol. 1990; 25: 678.

39. Jacobs C, Nicolay D, Grellet J, et al. effects of intravenous infusion of urographic contrast agents on glomerular filtration rate, serum concentration and urinary excretion of uric acid in subjects with normal renal function. Adv Exp Med Biol. 1987; 212: 145.

40. Cooper K, Bennett WM. Nephrotoxicity of common drugs used in clinical practice. Arch Intern Med. 1987; 147: 1213.

41. Harkonen S, Kjellstrand C. Contrast nephropathy. Am J Nephrol. 1981; 1: 69.

42. McCarthy CS, Becker JA. Multiple myeloma and contrast media. Radiology. 1992; 183: 519.

43. Banna M. Post-angiographic blindness in a patient with sickle cell disease. Invest Radiol. 1992; 27: 179.

44. Darr M, Hamburger S, Koprivica B, et al. Hemolytic anemia associated with a radiopaque contrast agent in a patient with haemoglobin SC disease. South Med J. 1981; 74: 1552.

45. Rao VM, Rao AK, Steiner RM, et al. The effect of ionic and non-ionic contrast media on the sickling phenomenon. Radiology. 1982; 144: 291.

46. Fradkin JE, Wolff J. Iodide-induced thyrotoxicosis. Medicine. 1983; 62: 1.

47. Salti IS, Kronfol NO. Aggravation of thyrotoxicosis by an iodinated contrast medium. Br J Radiol. 1977; 50: 670.

48. Shimura H, Takazawa K, Endo T, et al. T_4-thyroid storm after CT-scan with iodinated contrast medium. J Endocrinol Invest. 1990; 13: 73.

49. Gold RE, Wisinger BM, Geraci AR, et al. Hypertensive crisis as a result of adrenal venography in a patient with phaeochromocytoma. Radiology. 1972; 102: 579.

50. Raisanen J, Shapiro B, Glazer GM, et al. Plasma catecholamines in pheochromocytoma: effect of urographic contrast media. AJR. 1984; 143: 43.

51. Stanley P, Miller JH, Tonkin ILD, et al. Angiographic procedure. In: Stanley P, ed, Pediatric Angiography. Baltimore: Williams & Wilkins: 1982: 7.

52. Kuhn MJ, Baker MR. Optimization of low-osmolality contrast media for cranial CT: A dose comparison of two contrast agents. AJNR. 1990; 11: 847.

53. Ramsey RG, Czervionke L, Dommers M, et al. Safety and efficacy of sodium and meglumine ioxaglate (hexabrix) and hypaque M60% in contrast-enhanced computed cranial tomographic scanning: A double-blind clinical study. Invest Radiol. 1987; 22: 56.

This paper was first published in *Advances in X-Ray Contrast*. 1995;3:2–9.

UPDATE

A recent development in intravascular contrast agents is the government approval of usage of a new nonionic dimer, iodixanol, in the United States. This nonionic agent has 6 iodine atoms on each contrast molecule. Sodium, calcium and other electrolytes are included in the contrast material solution. Iodixanol is isotonic with blood at all available concentrations. The major advantage of this new agent compared with low osmolar media appears to be a further decrease in pain associated with intravascular injections. This is mainly advantageous for peripheral angiography, but it could be significantly better for some painful neuroangiographic procedures. Selective arteriography of the external carotid artery, or superselective arteriography of external carotid artery branches, can cause substantial patient discomfort. In selected cases this isotonic agent may be preferable. Radiographic efficacy with this new contrast agent is identical to other contrast materials. In the United States iodixanol is available in 2 iodine concentrations, 270 mg I/ml and 320 mg I/ml. This agent is slightly more viscous than other low osmolar agents and it should be heated to body temperature prior to injection to improve injectability. The dosage, injection rate, and filming sequence for this new agent should be identical to that used for standard intravascular iodinated contrast agents. The safety profile for this new iso-osmolar agent appears to be comparable to other low osmolar contrast agents.

Finally, it should be noted that this new iso-osmolar agent has not been approved for use in myelography in the United States.

P. Dawson and W. Clauss, (eds.), Advances in X-Ray Contrast: Collected Papers. 89–96.
© *1998 Kluwer Academic Publishers.*

Development of intravascular contrast agents: the first 100 years

Ronald G Grainger, MD, FRCP, FRCR, FRACR (Hon), FACR (Hon)
Kodak Professor of Diagnostic Radiology (emeritus), University of Sheffield, UK

WILHELM CONRAD ROENTGEN

Wilhelm Conrad Roentgen was born in Lennep, a small medieval textile town in Germany, in 1845, so that this year (1995) is not only the 100th anniversary of his discovery of X-rays on 8 November 1895, but also the 150th anniversary of his birth.

He was appointed Professor of Physics at the University of Würzburg in 1894. In the very next year, 1895, in his darkened laboratory he noticed, to his surprise, fluorescence of a barium platino-cyanide screen whenever he energized a nearby Hittorf gas discharge tube. For the next 2 months, he worked extremely hard on a very well-organized, meticulous scientific research programme, during which time he discovered and documented most of the major physical characteristics of his new, mystery X-rays.

He handed over his hand-written manuscript on 28 December 1895 and read out the presentation, to great acclaim, to The Würzburg Physical Medical Society on 23 January 1896. He published only three papers on X-rays (out of more than 60 of his publications) and these three papers are absolute 'classics' in precise and concise presentation of new scientific data and research. Most appropriately, he was awarded the First Nobel Prize for Physics in 1901. He donated his prize of 50 000 Swedish Kroner to the University of Würzburg, and as further evidence of his integrity and humanistic ethics, he refused to profit or take out any patents on his revolutionary discovery which changed medical practice for ever.

He died, aged 78 years, from bowel cancer and, most unfortunately and inexplicably, he demanded in his will that all of his laboratory and scientific notes and papers be destroyed by burning. But, fortunately, the most important historical paper, the first page of his hand-written account of his discovery of X-rays on 8 November 1895, escaped this wanton destruction and is still preserved today.

During Roentgen's original experiments at Würzburg University in November and December 1895, he demonstrated that materials of high atomic weight absorbed his new, mystery rays more effectively than did substances of lower atomic weight. He demonstrated this on the first human X-ray – that of his wife Berthe's hand – on 22 December 1895, on which her wedding ring is seen to be much more radio-opaque than the bones, which themselves are more radio-opaque than the skin and soft tissues.

Roentgen fully realized the significance of this X-ray plate. He took no less than 12 separate, 15-minute exposures of his wife's left hand and sent them, together with pre-prints of his first article on X-rays, to his physics colleagues and friends throughout the world.

The conventional history is that this was one of the first X-rays ever made, but in fact, 6 years earlier on the evening of 22 February 1890, at the University of Philadelphia, USA, two young researchers – Arthur W. Goodspeed (1860–1943), a physicist from New Hampshire, USA and William Nicholson Jennings (1860–1946), a scientific photographer born in Girlington, near Bradford, Yorkshire – inadvertently fogged some photographic plates, whilst discharging a Crookes gas discharge tube in a physics laboratory of Philadelphia University [1]. The canny Yorkshireman, Jennings, had put two coins for his trolley fare home on top of a pile of photographic plates. Similarly to my own disorganized laboratory practice, they were not sure which plates they had deliberately exposed, so they developed the entire stack of plates and were surprised to find that the top plate had been fogged and showed two round images of the coins. Like Alexander Fleming (40 years later), who established the new therapeutic era of antibiotics by preserving his culture plate contaminated by penicillin mould from the London smog, they retained this fogged and spoilt 1890 photographic plate. Six years later, after

Roentgen had published his new discovery in January 1896, they remembered and retrieved this spoilt plate and recognized it as an X-ray image. They did not challenge Roentgen for primacy of discovery: they praised him for his brilliant scientific research into the new rays, but they, I think reasonably, wished to record that in 1890, 6 years earlier, they had inadvertently made the first X-ray plate.

POST-MORTEM ANGIOGRAPHIC CONTRAST MEDIA

The first intentional use of radiological angiographic contrast medium was achieved a few days after 1 January 1896 when Roentgen, who was enjoying himself at Professor Exner's New Year Eve Party in Vienna, told the world of his Würzburg discovery [2]. A young physicist, Haschek, and a young physician, Lindenthal, injected a calcium carbonate emulsion into the brachial artery of the severed arm of a cadaver in Vienna. This first post-mortem arteriogram of the hand of the injected arm was of excellent quality and was given an exposure time of 57 minutes. A copy was sent to Roentgen. It was published on 23 January 1896 in *Klinische Wochenschrift* [3] (only 1 week after the plate had been taken).

The first successful visceral angiogram – a renal arteriogram – was probably achieved by Hicks, Professor of Physics at Sheffield University, on 6 February 1896 [4]. This post-mortem renal arteriogram, injected with red lead mass (as used in the dissecting room), was shown at the meeting of the Sheffield Medical Chirurgical Society on Thursday 13 February, 1896. The Sheffield doctors were so impressed that they started a charity fund to purchase a second X-ray machine for the Medical School and the Sheffield Royal Hospital, where I have had the privilege of working for the past 35 years. This renal arteriogram was published in the *British Medical Journal* on 22 February 1896 only 16 days after it had been made.

IODINE

Douglas Cameron, a young Minnesota surgeon, published in a preliminary report (in case he did not return from the World War) in *JAMA*, 1917, that oral and intravenous sodium iodide could produce a urinary cystogram opaque to X-rays [5]. This was confirmed when the Mayo Clinic team of Osborne (a 28-year-old syphilologist), Sutherland (radiologist), Scholl (urologist) and Rowntree (Professor of Medicine) published the first scientific paper in *JAMA*, 10 February 1923, on intravenous urography with intravenous sodium iodide [6]. This was after Osborne had noted (like Cameron) that the urinary bladder was radio-opaque on an abdominal radiograph of a syphilitic patient being treated with large oral and intravenous doses of sodium and potassium iodide. This Mayo Clinic team produced good quality intravenous cystograms and modest quality pyelograms and they related the administered iodide dose to the urinary concentration of iodine.

This paper established the potential of iodine as an intravascular contrast agent, but the problem was how best to package the enormous amounts of the toxic iodine atom required for organ and vascular X-ray imaging ($1\,000\,000\times$ the normal daily turnover of iodine may be required for a complex radiological study) into a safe, non-toxic molecule.

CLINICAL ANGIOGRAPHY

The first clinical human venograms were produced with inorganic solutions of strontium bromide in Frankfurt, Germany [7], and the first arteriograms with 100% sodium iodide in St Louis, USA [8].

During the mid-1920s, Egas Moniz (1874–1955), Professor of Neurology in Lisbon, Portugal, (a most remarkable and charismatic man of many skills – researcher, clinician, author, Foreign Secretary of Portugal, Member of Parliament before he qualified in Medicine in 1900) pursued the objective of clinical carotid arteriography for the diagnosis and localization of cerebral tumours. In a long series of animal experiments, he tried lipiodol emulsion, discarded the oil and then tried inorganic solutions of bromide and iodide salts of sodium, potassium, lithium, strontium and rubidium. His first successful human carotid arteriogram was obtained after injecting a 30% solution of sodium iodide into the surgically exposed carotid artery of a young man with a pituitary tumour, on 27 June 1927 [9]. He had previously tried percutaneous puncture of the carotid artery, but reverted to cut-down on the artery, using tapes temporarily to occlude the

91

artery.

Moniz was crippled by gout, especially affecting his hands, and he performed no injections himself, but he masterminded a very detailed research programme. Moniz was awarded the Nobel Prize for Medicine and Biology in 1949, but for the wrong reason – for prefrontal leucotomy for dementia, which technique he developed, but which is now virtually discredited and abandoned. In his citation for the Nobel Prize, there is no mention at all of his brilliant development of cerebral arteriography and of his charismatic leadership of a magnificent Portuguese team of young clinicians based at Santa Maria and Santa Marta Hospitals in Lisbon, who established clinical arteriography, aortography, venography, pulmonary angiography and lymphography between 1927 and 1932 [10].

Unfortunately, being disappointed with the toxicity, pain and poor radio-opacity of the inorganic halogen contrast agents, Moniz became a leading advocate of thorium arteriography, for thorium dioxide suspension was almost painless to inject, had very little acute toxicity and was densely radio-opaque, so that the patient could keep still during the long exposures necessary with the low power X-ray equipment. But thorium, being a very strong emitter of α radiation with a half-life of 10^{10} years, produced hundreds of malignant tumours (especially of the liver and biliary tract) 20–30 years later and is still the subject of medical litigation.

INTRAVENOUS UROGRAPHY (IVU)

Moses Swick (1900–1985), a 1924 graduate of Columbia University, New York, worked at Mount Sinai Hospital, New York, where he was awarded a Libman scholarship (Libman–Sacks DLE syndrome) to study research procedures abroad. He chose to work at the Altona medical clinic of Leopold Lichtwitz (1868–1943) in Hamburg, Germany, where he studied the therapeutic potential of a series of iodinated pyridine organic drugs synthesized by Arthur Binz (1868–1943) and Curt Rath, Professors of Chemistry of The Agricultural College in Berlin, with the objective of finding an improved magic bullet, superior to 'Salvarsan 606', to treat syphilis and other infections.

Swick observed that reasonable quality IVUs could

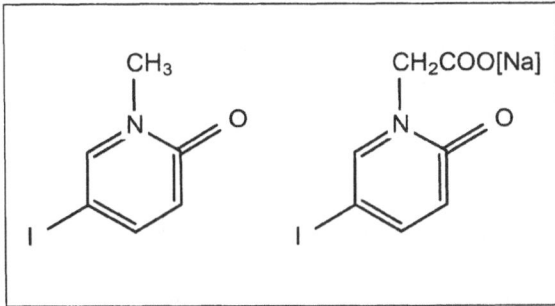

Figure 1 *Left:* Selectan Neutral: *N*-methyl-5-iodo-2-pyridone. *Right:* Uroselectan: 5-iodo-2-pyridone-*N*-acetate-sodium (1929). Redrawn from Br J Radiol 1982;55:1–18 with kind permission of the author

be obtained after intravenous injection of some of these iodinated organic pyridone chemicals into laboratory animals (mainly mice and rabbits), and he transferred his research to work on urological patients at the clinic of Professor Alexander von Lichtenberg (1880–1949) at St. Hedwig's Hospital, Berlin, which was the largest surgical urological clinic in the world. It was at that Berlin clinic, in 1929, that Swick succeeded in producing the first excellent quality diagnostic IVUs in several patients using Binz and Rath preparations – at first with 'Selectan Neutral' (a non-ionic, mono-iodinated pyridone molecule) and later with the less toxic, but more soluble, 'Uroselectan' (sodium 5-iodo-2-pyridone-*N*-acetate; patent rights awarded to Rath in Berlin on 12 May 1927) (Figure 1).

In 1929, von Lichtenberg (who had introduced 'Kollargol' (colloidal silver) for retrograde pyelography in 1906) was on the inevitable professorial extended lecture tour in America (that practice has not changed up to 1995), whilst Swick was doing the real work in the development of the first IVUs in Lichtenberg's Department in Berlin. Tragically, a bitter personal confrontation developed between the young, inexperienced American, Swick, and the prestigious German (Hungarian-born) professor, concerning attribution of the successful research and the priority of publication. It was eventually decided that the first paper reporting the successful research of the very important (even sensational) diagnostic discovery

92

should be by Swick alone [11].

Like Roentgen before him, Swick refused to benefit financially from his dramatic discovery of IVUs and he donated his $1500 award from Schering Kahlbaum (Berlin pharmaceutical company) to The Libman Foundation, which had originally funded his research work in Germany. The world of medicine, however, believed that von Lichtenberg, and not Swick, was the real discoverer of IVUs.

But Professor Victor F. Marshall (then Professor of Urological Surgery in New York) conducted a very meticulous investigation of all of the evidence for the discovery of IVU and concluded "Swick's contribution shines through the dust of the scramble for priorities" [12].

It took no less than 37 years for Swick to be recognized. On 17 March 1965, Swick was awarded the Valentine Award of the New York Academy of Medicine, and he was introduced by the chairman (Myer Melicow) thus: "And now 30 years have passed, 30 unkind years of heartache and oblivion".

At this meeting of the Academy of Medicine, the great American urologists publicly apologized to Swick, and this was followed by the award of many international honours and distinctions, including the American Congressional Record (16 May 1978): "One of the five major contributions of an individual in Medicine". But after a medical life-time (37 years) of destructive criticism and allegations of cheating and plagiarism, ostracized by many of his colleagues and by the international medical establishment, Swick was understandably still very upset when this author (Grainger) managed, with difficulty, to trace and meet him at his home in New York in 1980.

The November 1929 issue of *Klinische Wochenschrift* (which printed Swick's original IVU paper) [11] was a wonderful issue, for the immediately preceding article is the very first paper by Werner Forssmann (1904–1979) (aged only 25 years and in his first house job in the very small Augusta Victoria Hospital in the little town of Eberswalde, near Berlin), in which he reported and illustrated the first catheterization of the human heart [13]. He was the catheterizer and he was also the very brave and (rather) crazy patient who became obsessed with dangerous self-experimentation (9 cardiac self-catheterizations, attempted angiocardiography with sodium iodide solution, 16 vein cut-downs, attempted direct puncture aortography) [14].

Forssmann was prohibited from further hospital work ("Tricks for the circus, not for a decent University Hospital" (Sauerbruch – the famous thoracic surgeon)). His pioneer cardiac catheterization work, however, was developed several years later by Cournand and Richards, in America, and McMichael, in London. After he had shared the 1956 Nobel Prize for Medicine with Cournand and Richards, he was at last recognized and honoured in his native Germany and became Professor of Surgery!

Forssmann's pioneer catheterization led directly to Sven I. Seldinger's first paper on percutaneous catheterization in 1953 [15] (so impressed were his chiefs at the Karolinska Hospital in Stockholm that they would not give him one day's leave to read the paper at Helsinki; and that wonderful paper was presented by title only). This again shows that our senior establishment do not always make the correct decisions! That paper by Seldinger (unappreciated by his colleagues and seniors) opened the way for the whole of the radiological practice of selective angiography and interventional radiology.

In the early 1930s, Uroselectan was superseded by two improved iodinated pyridone products – iodoxyl (Uroselectan B) and iodopyracet (Diodone) – both containing two atoms of iodine, and both synthesized by the Binz and Rath team in Berlin, who were supported in their research by Schering Kahlbaum. Berlin was certainly the innovative centre for contrast medium research in 1927–30.

BENZOIC ACID DERIVATIVES

The substituted benzene ring was introduced in 1933 as a contrast agent by Moses Swick (who persisted in his IVU research, despite the lack of acknowledgement mentioned above) and Vernon Wallingford (b. 1896) of Mallinckrodt Chemical Works (USA), but their molecule, sodium monoiodohippurate, containing only one atom of iodine, was rather toxic (the iodine atom reduces the LD_{50} of hippuric acid by 50%) and was not as satisfactory as iodoxyl and iodopyracet. In the early 1950s, the same Wallingford [16] made a major discovery when he showed that the introduction of an amine group in the meta- (C3) position of the benzene ring allowed the introduction of three atoms of iodine

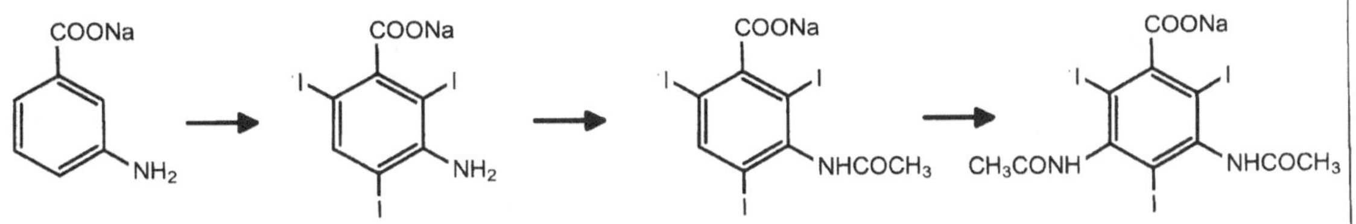

LD$_{50}$: 20   1.5   15.7   22
        A   B   Sodium acetrizoate   Sodium diatrizoate
                 (Urokon: Diaginol)   (Urografin: Hypaque)

(after Hoppe et al, 1956 [17])

Figure 2 Addition of amine group (–NH$_2$) to sodium benzoate (A) permits tri-iodination at C2, C4 and C6 (B), but LD$_{50}$ is reduced from 20 to 1.5. This product is far too toxic. Acetylation of amine group (sodium acetrizoate; Diaginol) increases LD$_{50}$ to 15.7. Addition of a second amino-acetyl group at C5 (sodium diatrizoate; Hypaque, Urografin) further increases LD$_{50}$ to 22. The higher the LD$_{50}$, the less the toxicity and the greater the clinical tolerance. [Redrawn from Br J Radiol 1982;55:1–18 with kind permission of author]

(C2, 4, 6 positions) (Figure 2), but this reduced the LD$_{50}$ by no less than 14 times and rendered the product far too toxic! However, he showed that acetylation of the amine group greatly reduced the toxicity, increasing the LD$_{50}$ of the tri-iodinated molecule by as much as 10 times, and the resulting product – sodium acetrizoate – was introduced in the early 1950s as a very useful contrast agent ('Urokon').

Hoppe et al [17] from the USA and other researchers in Europe introduced a second acetylamino group at C5 which further reduced the toxicity, increasing the LD$_{50}$ by a further 50%, to regain the low toxicity of the non-iodinated amino-benzoate.

In the mid-1950s the very much improved sodium and meglumine salts of the above, fully-substituted tri-iodinated benzoic acids (diatrizoic and iothalamic acids) with three atoms of iodine and two acetylamino groups attached to one benzene ring – sometimes known as 'Hypaque', 'Urographin' and 'Conray' – were introduced into clinical practice. They became the standard contrast agents of choice for the next 25 years until the 1980s and they are still being used today. However, they are very hyperosmolar, being 5–8 times the osmolality of every cell in the body, and this hyperosmolality greatly increases their haemodynamic and other toxicity parameters.

The next major development occurred in 1969,

when a young Swedish radiologist, Torsten Almén (b. 1931), self-taught in organic chemistry, postulated that the high osmolality of these tri-iodinated benzoic acid ionizing salts could be greatly reduced by transforming the salt molecule into a non-ionizing product such as an amide. Almén's original paper, entirely theoretical with virtually no chemical or clinical research, was understandably rejected by several major radiological journals (who, thereby, missed the most important contrast medium paper in 50 years!). He eventually succeeded in publishing in *The Journal of Theoretical Biology* (1969) [18] – a journal the existence of which most radiologists were, and still are, completely ignorant!

Almén's pioneering approach and continuing research into contrast media were justly recognized by the award to him of the 1987 Fernstrom Great Nordic Prize (a miniature Nobel Prize for Medicine). He was the first radiologist to be so honoured and he is probably the only clinical radiologist to have made a major breakthrough in the development of contrast agents.

Almén's ideas of replacing the ionizing salt molecules by non-ionizing and, therefore, low osmolar amides were developed commercially by Nycomed of Oslo, with which company he has been closely associated ever since. Metrizamide, the first tri-

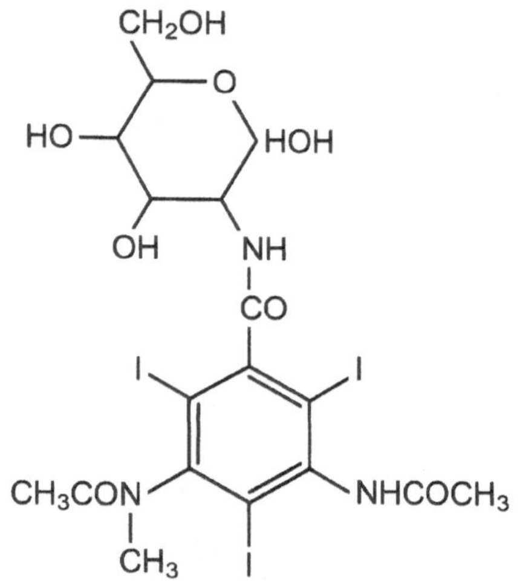

Figure 3 Metrizamide ('Amipaque'). Note substituted amide (CONH$_2$) group at C1 instead of COONa in sodium diatrizoate (Figure 2). No ionization – therefore low osmolality. [Redrawn from Br J Radiol 1982;55:1–18 with kind permission of author]

iodinated non-ionic, low-osmolar contrast medium (LOCM) so developed, was marketed in the early 1970s, mainly as a myelographic agent (Figure 3). It was, however, unstable in solution and the lyophilized powder had to be dissolved immediately before injection. It was about 20–30 times more expensive than conventional high-osmolar contrast media (HOCM) ('Conray', 'Urographin') and was too expensive for general intravascular use.

Fortunately, the pharmaceutical manufacturers (particularly Bracco, Guerbet, Mallinckrodt, Nycomed and Schering) soon greatly improved on the original metrizamide and synthesized and marketed the less toxic, more stable and much less expensive second generation LOCM. The first batch of these, the non-ionic iohexol ('Omnipaque') and iopamidol ('Nio-pam') and the ionic ioxaglate ('Hexabrix') were introduced in the mid-1980s and are still among the contrast media of choice today, 10 years later (Table 1). In the last few years, several new non-ionic, low

osmolar contrast media have been synthesized and marketed. They have interesting, useful but relatively minor advantages over iohexol and iopamidol, but they extend the patents for the pharmaceutical companies.

THE FUTURE

Recent developments have also included the synthesis of the non-ionic dimers (iodixanol, iotrolan), each molecule of which contains six atoms of iodine, and which are isotonic with body tissues at all concentrations. These non-ionic dimers probably have some advantages for intrathecal and intravascular use, but the gain is partly countered by the increased viscosity and expense and the possibility of increased renal toxicity.

Our current low-osmolar, tri-iodinated, radiological contrast agents are now of such high quality and safety that our major requirement in this decade is for lower-cost media with the same efficiency and low toxicity. We also need long-stay blood-pool imaging agents, particularly for CT.

The world market for radiological iodinated contrast media for conventional radiology and CT is approximately 3 billion dollars in 1995, with about 70% volume usage of the more expensive, low-osmolar media which account for about 85% of the expenditure.

However, the main thrust of contrast media research in the next few years will almost certainly be in the non-radiological imaging systems, MRI and ultrasonics, both of which can image the vessels well, even without added contrast agents. But in order to improve vessel imaging, research into non-radiological contrast media is proceeding at great pace. For example, it is forecast that the market in the USA alone for ultrasonic media (for perfusion agents, chamber opacification and Doppler imaging) will grow from the present zero to 1 billion dollars in the next 5 years.

Gadolinium chelates have already established a considerable and increasing usage for MRI, and different paramagnetic agents are being actively researched. Some are now being introduced into clinical practice. It is forecast that the MRI contrast agent market may reach 1.5 billion dollars in the next 5 years.

This has been a wonderful 100 years of imaging agent development in various imaging modalities, of which we should be very proud. Roentgen himself

Table 1. Iodinated contrast media

Classification	Iodine atoms per molecule	Osmotic particles per molecule	Iodine:particle ratio	Molecular weight	Iodine content at 0.3 osmol/kg water (approx.)	Osmolality at 300 mg I/ml (osmol/kg water) (approx.)	LD_{50} g I/kg wt mouse at 2 ml/min i.v. (very approx.)
Ionic monomer Diatrizoate* Iothalamate* Metrizoate Ioxithalamate Iodamide Ioglicate	3	2	3:2	600–800	70	1.5–1.7	7
Non-ionic monomer Iopamidol* Iohexol* Ioversol* Iopentol Iopromide** Ioxilan**	3	1	3:1	600–800	150	0.6–0.7	22
Ionic dimer Ioxaglate*	6	2	3:1	1269	150	0.56	12
Non-ionic dimer Iodixanol** Iotrolan	6	1	6:1	1550–1626	300	0.30	⩾ 26

From Grainger RG and Allison DJ. Textbook of Organ Imaging. An Anglo-American Text, 3rd edn. Edinburgh and London: Churchill-Livingstone; 1996; in press. Modified from Morris TW [19]

* Already available in the US
** Currently in US trials

would surely be very impressed and perhaps even amazed at these unexpected developments in contrast medium for vessel and organ imaging, both by his own X-rays and by the new, non-radiological imaging systems developed in the last 20 years. He initiated this continuing and remarkable development of imaging by recognizing the enormous potential of that complete accident on 8 November 1895 in his laboratory in Würzburg, 100 years ago.

This paper is presented with permission and based on a lecture delivered by the author at The Birmingham Roentgen Centenary meeting on 14 June 1995 and published in Thomas AMK, editor. The Invisible Light. Oxford: Blackwell Science; 1995. A more extensive and detailed, well-illustrated account and bibliography with photographs of many of the personalities mentioned above is provided in: Grainger RG. Intravascular radiology – the past, the present and the future. Br J Radiol. 1982; 55: 1–18.

REFERENCES

1. Walden TL Jr. The first radiation accident in America: A centennial account of the X-ray photograph made in 1890. Radiology. 1991; 181: 635–639.
2. Doby T. Development of angiography and cardiac catheterization. Littleton, MA: Publishing Sciences Group; 1976.
3. Haschek E, Lindenthal TO. Ein Beitrag zur Praktischen Verwerthung der Photographie nach Röntgen. Wien Klin Wochenschr. 1896; 9: 63–64.
4. Rowlands S. Report on the application of the new photography to medicine and surgery. Br Med J. 1896; Feb 22: 492–497.
5. Cameron D. Aqueous solutions of potassium and sodium iodides as opaque media in roentgenography. JAMA. 1917; 70: 754–755.
6. Osborne ED, Sutherland CG, Scholl AJ Jr, Rowntree LG. Roentgens of the urinary tract during excretion of sodium iodide. JAMA. 1923; 80: 368–373.
7. Berberich J, Hirsch SR. Die Roentgenographische Darstellung der Arterien und Venen am lebenden Menschen. Klin Wochenschr. 1923: 2226–2228.

8. Brooks B. Intra-arterial injection of sodium iodide. JAMA. 1923; 82: 1016–1019.

9. Moniz E. Arterial encephalography. Its importance in the location of cerebral tumours. Rev Neurol. 1927; 48: 72.

10. Veiga-Pires JA, Grainger RG. Pioneers in angiography: the Portuguese school of angiography. Lancaster, Boston, The Hague: MTP Press; 1982.

11. Swick M. Darstellung der Niere und Harnwege in Roentgenbild durch intravenose Einbringung eines neuen Kontraststoffes: des Uroselectans. Klin Wochenschr, 1929; 8: 2087–2089.

12. Marshall VF. The controversial history of excretory urography. In: Emmett J, editor. Clinical urography, Vol. 1. Philadelphia: Saunders; 1977: 2–5.

13. Forssmann W. Die Sondierung des rechten Herzens. Klin Wochenschr. 1929; 8: 2085–2087.

14. Forssmann W. Experiments on myself. Memoirs of a surgeon in Germany. New York: Saint Martins Press; 1972.

15. Seldinger SI. Catheter replacement of the needle in percutaneous arteriography; a new technique. Acta Radiol. 1953; 39: 368.

16. Wallingford VH. The development of organic iodide componds as X-ray contrast media. J Am Pharmacol Assoc (Scientific Edition). 1953; 42: 721–728.

17. Hoppe JO, Larsen HA, Coulston FJ. Observations on the toxicity of a new urographic contrast medium, sodium 3,5-diacetamido-2,4,6-tri-iodobenzoate (Hypaque sodium) and related compounds. J Pharmacol Exp Ther. 1956; 116: 394–403.

18. Almén T. Contrast agent design. Some aspects on the synthesis of water-soluble contrast agents of low osmolality. J Theoret Biol. 1969; 24: 216–226.

19. Morris TW. X-ray contrast media. Where are we now and where are we going? Radiology. 1993; 188: 11–16.

This paper was first published in *Advances in X-Ray Contrast.* 1995;3:26–33.

UPDATE AND COMMENTS

There have been no major developments in our knowledge of contrast media since the above paper was written and published.

There has been increasing and continuing development in recording of blood vessels and organs by the newer imaging modalities of CT, MRI, ultrasound, and Doppler. New imaging sequences have considerably enhanced magnetic resonance angiography in the last few years. It will be interesting to observe the increasing competition between conventional radiographic imaging, magnetic resonance angiography, spiral/helical CT, ultrasound and Doppler (with or without their specific contrast media) as we approach the millennium.

P. Dawson and W. Clauss, (eds.), Advances in X-Ray Contrast: Collected Papers. 97–106
© *1998 Kluwer Academic Publishers.*

Risk factors, prophylaxis and therapy of X-ray contrast media reactions

William H Bush Jr, MD, FACR
University of Washington School of Medicine, Department of Radiology, Box 357115, 1959 NE Pacific Street, Seattle, WA 98195, USA

INTRODUCTION

Although the radiocontrast media (RCM) used routinely for the past 30 years are relatively safe drugs, adverse reactions to these conventional, ionic contrast agents can be expected with an overall frequency of 5–12% [1–5]. In one out of every 1000 examinations, a reaction of life-threatening severity may occur [6,7].

Newer radiocontrast agents are more physiological because of their lower osmolality; most are non-ionic. The relative risk of adverse reactions to these lower osmolality, non-ionic media is a factor of five less than with conventional, higher osmolality ionic agents, but reactions, even fatal reactions, are not totally eliminated [3,8–11].

AETIOLOGY OF ADVERSE REACTIONS

Adverse reactions to intravascular RCM are generally classified as either idiosyncratic (pseudoallergic) or chemotoxic. Idiosyncratic (i.e. pseudoallergic, anaphylaxis-like, allergic-like, anaphylactoid) reactions to intravascular RCM occur unpredictably and independently of the dose or concentration of the agent. The exact cause of these allergic-like reactions to RCM remains uncertain [12,13]. As noted by Greenberger [13], there are arguments for and against an immunological mechanism: reactions simulate an IgE-mediated process, yet contrast media do not seem to react with tissue proteins to form immunogens; reaction prevalence is increased in previous reactors, yet a reaction may occur on initial exposure to a contrast medium and previous reactors may not react to subsequent exposure to the same agent. Specific antibodies to the responsible contrast medium are seldom detected; Brasch et al [14] are essentially alone in being able to demonstrate antibodies to radiocontrast agents. The vast majority of investigators dispute adverse reactions to contrast media being IgE-mediated reactions [13,15,16].

Contrast media can cause the direct release of histamine from mast cells and basophils, and systemic histamine produces reactions similar to those observed after the intravascular administration of RCM [17,18]. Subclinical bronchospasm occurs routinely after intravenous RCM for excretory urography [19]. Contrast media can activate directly or indirectly the complement, coagulation, fibrinolytic and kinin systems leading to the release of multiple mediators (histamine, leukotrienes, lysosomal enzymes, fibrin-split products, bradykinin) capable of producing the adverse affects noted after injection of contrast media [16,20–24]. However, the degree to which these systems are activated by a contrast medium does not correlate with observable reactions. In idiosyncratic, anaphylactoid reactions, an important consideration seems to be the extent to which the aforementioned systems are perturbed at the time the contrast medium is injected [4,21]. Radiocontrast inhibition of cholinesterase has also been proposed as a mechanism, with enhanced cholinergic activity, causing vasodilatation and increased blood flow, bronchospasm, urticaria, cardiac rhythm disturbances and convulsions [25]. The lower osmolality, non-ionic agents appear to inhibit cholinesterase less than do conventional, ionic media.

Contrast reactions resulting in severe acute cardiovascular collapse or cardiopulmonary arrest, without cutaneous or bronchospastic components, make it impossible to disregard Lalli's theory [26] that exposure of the 'bare' or unprotected areas of the central brain is the mechanism by which some, if not all, of the manifestations of contrast reactions occur. Animal studies [27,28] on the effect of RCM on the blood–brain barrier add some theoretical support to Lalli's hypothesis.

Chemotoxic reactions to RCM result from the physicochemical effects of the medium on the organs or vessels. Unlike idiosyncratic reactions, chemotoxic effects are directly dependent on the dose and concentration of the administered agent. Hence, the rate and site of injection play important roles in the intensity and nature of the resulting effects. The contrast medium's potential for binding calcium ions and its capacity for hydrophobic interactions with biological molecules, together with the nature and concentration of the cations (i.e. sodium or meglumine) of the ionic agents, cause the chemotoxic effects [4].

PREVALENCE OF REACTIONS

Most acute adverse reactions are minor, and no treatment is required. A non-life-threatening, moderate reaction requiring some treatment occurs in 1–2% of patients receiving conventional ionic radiocontrast [29]. Severe, life-threatening reactions can be expected in 0.06–0.4% (pooled data 0.13%) of patients receiving conventional, ionic RCM. The relative risk of having a reaction to the lower osmolality, non-ionic RCM is at least a factor of five lower than the risk with a conventional, ionic agent [3,11,30–34].

Reported mortality rates have varied from one in 13 000 [35] to one in 169 000 [3]; a commonly quoted mortality rate is one in 75 000 [2]. The most recent study [3] had a low mortality rate (1:169 000) that was equal for both ionic and non-ionic radiocontrast agents. The overall decrease in mortality over the past two decades reflects many factors, especially the education of personnel in the treatment of reactions and the routine use of lower osmolar, non-ionic RCM for those patients at higher risk for an adverse effect [3,29,36]. A cautionary note: a severe, life-threatening, anaphylaxis-like reaction can occur with even a small dose of lower osmolality, non-ionic RCM [9,10]. In my own experience, a patient with a history of only mild asthma received less than 5 ml of lower osmolality, non-ionic RCM intravenously and experienced severe bronchospasm and hypotension.

The overall prevalence of reactions to the gadolinium-based contrast agents for magnetic resonance (MR) imaging is approximately 1–2%; severe systemic anaphylaxis-like reactions to these agents occur rarely. The occurrence of a severe reaction is only 1:350 000 injections, and asthma appears to be a risk factor [37].

Delayed reactions of a systemic nature to RCM are much more common than previously appreciated [38]. Patients report a similar frequency (approximately 27%) and nature of reactions to both ionic and non-ionic RCM [39]; most were not serious or life-threatening: 15% had a 'flu-like illness, 1% parotitis, 5% nausea/vomiting/abdominal pain, and 3% headache.

RISK OF REACTIONS

Age: Overall, adverse reactions appear to be most frequent for persons 20–50 years old, less frequent for persons older than 50 years, and even less frequent for persons younger than 20 years. Reactions to RCM may be most severe in older persons, who are unable to withstand a severe systemic reaction, especially with cardiopulmonary manifestations. Furthermore, elderly patients are more susceptible to the chemotoxic effects of contrast media.

Allergy and Asthma: Persons with allergic tendencies are at increased risk for an idiosyncratic reaction. If there is a history of systemic allergies to multiple substances, the relative risk of reaction to RCM is approximately twice that for the general population; for patients with a history of asthma, this relative risk is approximately five times greater [40,41].

Medications: There is debate as to whether persons taking β-adrenergic blocking drugs have an increased risk of anaphylactoid reactions. In a study by Greenberger [42], anaphylactoid reactions were not more frequent when the patient was using cardioselective β-blockers, non-selective β-blockers, or calcium antagonists; however, there was a trend to more reactions if the patient was on β-blockers. Patients taking calcium antagonists, which prevent IgE-mediator release from lung tissue, did not have a reduced prevalence of bronchospastic-type reactions. Conversely, in a study by Lang et al [40], there was a statistically significant increase in the frequency of anaphylactoid reactions in patients taking β-adrenergic blocking drugs, even an ophthalmic preparation. Both groups noted that response to treatment for contrast reactions was 'sluggish' or 'refractory' in the patients taking β-blockers.

As best as one can determine, there is no direct correlation or association between 'allergy' to povidone-iodine skin cleansing solution and the predisposition to an allergic-like reaction to intravascular RCM. The former is considered a contact dermatitis, the latter an idiosyncratic reaction.

Of interest, is the increased prevalence of 'recall reactions' after administration of RCM to patients receiving interleukin-2 (IL-2), a potent stimulant of the human immune system [43,44]. More than 10% of patients who received recombinant interleukin-2 plus intravenous RCM developed delayed reactions resembling those occurring after systemic IL-2 therapy; fever, chills, malaise, nausea and vomiting, skin rash, diarrhoea and occasionally hypotension occurred on re-exposure to intravenous RCM for follow-up examination performed 1 month after the initial IL-2 and RCM administration [43]. The reactions responded promptly to supportive therapy. In the study by Shulman et al [45], interleukin-2 'recall' reactions occurred less often with non-ionic contrast media; reaction prevalence was also reduced by waiting more than 4 weeks (i.e. 4–6 weeks) between IL-2 treatment and intravenous RCM administration. Different reactions, resembling hypersensitivity reactions to RCM, occurred in patients who received intra-arterial ionic radiocontrast followed by intra-arterial infusions of IL-2; on re-exposure to RCM 1 month later, more than 25% of patients experienced an anaphylactoid reaction [46]. These reactions were reversed by corticosteroids, and prevented subsequently by a corticosteroid premedication regimen. However, corticosteroids may block IL-2 effector mechanisms, so routine corticosteroid premedication for RCM administration is not recommended [45]. Corticosteroids should be reserved to treat a serious RCM reaction if one occurs despite the precautionary steps of establishing an adequate interval between IL-2 and RCM, the use of a non-ionic medium, and initial treatment of a reaction with supportive measures.

The biguanides, phenformin and metformin, are used as monotherapy or combination therapy for patients with non-insulin-dependent diabetes mellitus (NIDDM). Renal insufficiency and liver failure are contraindications to the use of these drugs because failure of renal excretion of metformin or failure of hepatic metabolism and excretion of phenformin leads to accumulation of these biguanides and the potential for fatal lactic acidosis [47]. The use of RCM in patients receiving metformin is currently very controversial. The potential danger of lactic acidosis exists because of accumulated metformin (i.e. from failure of renal excretion) and not from any interaction of RCM and metformin. Patients at risk are those who may develop RCM-induced nephrotoxicity and renal insufficiency/failure. It is currently advised that metformin be stopped 48 hours before, and for 48 hours subsequent to, the administration of RCM; normal renal function should be confirmed before metformin is resumed. The unresolved controversy is whether metformin need be stopped before RCM is administered to patients with entirely normal renal and liver function.

Total dose of RCM: The total dose probably affects the prevalence of chemotoxic-type reactions to RCM but will have little influence on idiosyncratic reactions. This may account for the disparity in the literature about whether the occurrence of clinically significant adverse reactions is related to dose [41].

Prior reaction to RCM: The prevalence of adverse reactions to conventional, ionic RCM in patients who have had a previous reaction to a conventional RCM agent and who have not received pretreatment with corticosteroids, is 17–35% [48,49], or three to eight times, perhaps even 11 times, greater than the risk for the general population. Administration of lower osmolality, non-ionic RCM to such patients reduces the frequency of repeated reactions to 5% [34,50].

PREVENTION OF ADVERSE REACTIONS

Chemotoxic reactions

Chemotoxic effects of RCM are more likely in debilitated or medically unstable patients. Hence, patients should be screened for conditions such as renal dysfunction, renovascular disease, severe cardiovascular disease, or recent seizures. Alternative diagnostic procedures that do not require RCM can

100

be considered. If RCM are necessary, a lower osmolality, non-ionic agent should be used. Gadolinium agents, because of their greatly reduced reaction prevalence, can be used during MR imaging of those patients who have had a previous reaction to RCM or who have renal dysfunction.

All patients should be well hydrated before intravascular injection of any RCM. In patients with renal insufficiency, nephrotoxic effects of RCM can best be mitigated by active hydration (e.g. 0.45% saline) i.v. [51]. For patients with diabetes and renal dysfunction, lower osmolar, non-ionic RCM offer some, but not much, lessening of nephrotoxic potential [52,53]; the reduction in nephrotoxicity is certainly not of the same order of magnitude as that which these non-ionic agents offer in the reduction of allergic-like reactions.

Anaphylactoid reactions

Pretesting (intravascular, cutaneous) of patients to detect those who have an increased likelihood of having an anaphylaxis-like reaction to RCM has been abandoned because it is insensitive and, in itself, potentially dangerous [6,54,55].

Patients who have asthma or multiple systemic allergies should receive lower osmolality, non-ionic RCM. If a patient had a mild prior reaction, the use of non-ionic RCM, without additional premedication, is considered sufficient if the patient does not have other risk factors [11,34,50]. If a patient had a previous moderate reaction to conventional, ionic RCM, pretreatment with corticosteroids and antihistamines (i.e. H_1 histamine blocker, such as diphenhydramine), plus using a non-ionic, lower osmolality RCM, is recommended [56–61]. Antihistamines alone (without concomitant corticosteroids) have not proved effective as a premedication regimen [6]. If a patient previously had a severe reaction to RCM, alternative diagnostic procedures that do not require RCM are advised. An option for some diagnostic procedures is gadolinium-enhanced MRI.

Pretreatment regimens: Kelly et al [58] used a corticosteroid (prednisone) and an H_1 antihistamine (diphenhydramine) and found that the prevalence of repeat reactions to conventional, ionic RCM decreased in their patients, from historical numbers of 17–35% to

Table 1   Pretreatment regimens

Option A (corticosteroid, H_1 antihistamine, non-ionic contrast medium) [57]:
   Prednisone 50 mg p.o., 13 h, 7 h, 1 h before contrast medium
   Diphenhydramine 50 mg p.o., 1h before contrast medium
   Non-ionic, lower-osmolality contrast medium

Option B (corticosteroid, non-ionic contrast medium) [59]:
   Methylprednisolone 32 mg p.o., 12 h and 2 h before contrast medium
   Non-ionic, lower-osmolality contrast medium

approximately 5% (Table 1). Using this same corticosteroid–antihistamine regimen, Greenberger et al [62] found that repeat reactions occurred with an incidence of 11%. With the addition of ephedrine sulphate to the corticosteroid–antihistamine regimen, reactions in the Greenberger study decreased to 5%, whereas the inclusion of cimetidine, an H_2 histamine blocker, increased the overall occurrence of reactions to 14%. Of note was increased tremulousness in patients who received ephedrine; furthermore, ephedrine is inadvisable in patients with a history of hypertension or cardiovascular disease. More recently, Greenberger and Patterson [57] found a significant further reduction (i.e. to 0.5%) in the occurrence of anaphylactoid reactions in previous contrast reactors when they received the corticosteroid–antihistamine pretreatment regimen and were examined using lower osmolality, non-ionic RCM. A similar decrease with non-ionic RCM was also shown in a prospective, randomized study by Lasser et al [59].

In a large, prospective study, Lasser et al [60,63] investigated the potential benefit of pretreating all patients with an oral corticosteroid (methylprednisolone, 32 mg) given 12 h and 2 h before conventional, ionic RCM. Significant reductions in the overall number of reactions and in mild reactions were noted. There was a trend, though not statistically significant, towards a reduction in the number of reactions requiring treatment and in the number of severe reactions, but reactions were not totally eliminated. Other authors disagree that routine steroid prophylaxis is advantageous [64]. Moreover, although corticosteroid–antihistamine pretreatment appears beneficial in reducing the occurrence of anaphylactoid reactions, it has little effect on chemotoxic reactions. Additionally, in children with acute lymphoblastic leukaemia and non-Hodgkin lymphoma, a tumour lysis syndrome is of sufficient concern that use of corticosteroids as prophylaxis against potential reactions to RCM should be avoided [65]. However, for many patients, the approach of routine administration of corticosteroids may be an alternative to routine use of the expensive lower osmolality, non-ionic RCM; issues of patient compliance and practicability do arise.

Typical adverse reactions to RCM

A vast array of reactions may occur, either singly or in combination:

1. Nausea/vomiting

2. Cutaneous reactions of scattered hives to extensive urticaria (yet no respiratory symptoms)

3. Airway reactions: bronchospasm, laryngoedema (without coexisting cutaneous or vascular effects)

4. Hypotension, with normal sinus rhythm or tachycardia

5. Vagal reaction (hypotension and sinus bradycardia)

6. Anaphylaxis-like reaction (rapidly progressive or severe bronchospasm, angioedema, laryngospasm, urticaria, rash, hypotension)

7. Seizure

8. Cardiovascular collapse, cardiac arrest

Treatment for specific reactions

Table 2 and the following discussion summarize specific treatment plans for the more frequently occurring adverse reactions [36,66]. However, these recommendations are not the only effective treatments [67–69], and physicians are encouraged to develop their own/departmental protocols, which should be reviewed and updated periodically.

Nausea and vomiting

Nausea and vomiting, though usually self-limited, may be the first signs of a more severe reaction. Lalli found that 15% of fatal reactions with urography began with nausea and vomiting [26]. For this reason, the patient should be observed closely for systemic symptoms, while maintaining intravenous access. The rate of injection should be slowed or stopped; changing RCM to a lower-osmolality, non-ionic agent often mitigates nausea and vomiting.

Cutaneous reactions

Treatment is usually not necessary for only a few scattered hives or pruritis. However, the patient should be observed closely for other developing systemic symptoms, while maintaining intravenous access. Urticaria that is extensive or bothersome to the patient is treated with an H_1 antihistamine (e.g. diphenhydramine). Profound cutaneous reactions may respond better to the addition of an H_2 antagonist (e.g. cimetidine, ranitidine or rizatidine). With diffuse erythema and urticaria, 'third spacing' of fluid and hypotension may occur; in such situations, treatment with i.v. fluids (e.g. normal saline) and i.v. epinephrine is recommended.

Airway reactions (bronchospasm and laryngeal oedema)

Bronchospasm, without co-existing cardiovascular problems, should be treated with oxygen and inhaled bronchodilators. Inhaled β_2-adrenergic agonists (e.g. albuterol, metaproterenol or terbutaline) can be delivered in high dose directly to the airways and have minimal systemic cardiovascular effects. The usual dosage, using a metered dose inhaler (MDI), typically involves 2–3 deep inhalations. Aminophylline is no longer a 'first-line' drug, having been supplanted by the β_2-agonist inhalers; it may also cause significant hypotension. I no longer recommend its use; instead, if the initial β_2-agonist inhaler is not fully effective, add nebulized β_2-agonist medication, or inhaled mist, or use i.v. epinephrine. Conversely, laryngospasm and laryngeal oedema usually do not respond to inhaled β_2-agonists, and these agents may worsen laryngeal oedema. Therefore, clinical evaluation and auscultation of the patient prior to beginning treatment is extremely important.

Epinephrine is the primary treatment for laryngeal oedema and is indicated for bronchospasm unrelieved

Table 2 **Acute reactions to contrast media: Treatment outline**

Urticaria:
Mild: observation
H_1 antihistamine (e.g. diphenhydramine 25–50 mg
p.o./i.m./i.v.

Severe: **add:**
epinephrine (1:10,000), 1 ml (0.1 mg), i.v. slowly (e.g. over 2–5
min)
H_2 antihistamine:
e.g. cimetidine injectable, 300 mg, diluted to 20 ml, slowly i.v.
(paediatric: 5–10 mg/kg, diluted to 20 ml, slow i.v.)
. . . *or* . . .
e.g. ranitidine injectable, 50 mg, diluted to 20 ml, i.v. slowly
(paediatric: use not established)
i.v. fluids

Bronchospasm (isolated):
Oxygen (6–10 L/min)
β_2-Agonist metered dose inhaler (2–3 deep inhalations): (e.g.
metaproterenol, terbutaline or albuterol) (or use nebulizer if
available)
Epinephrine
Subcut: 1:1000, 0.1–0.2 ml (0.1–0.2 mg)
(paediatric: 0.1–0.2 mg subcut)
i.v.: 1:10,000, 1 ml (0.1 mg), slowly (e.g. over 2–5 min)
(paediatric: 0.01 mg/kg, i.v.)

Hypotension (isolated):
Oxygen (6–10 L/min)
Elevate patient's legs
i.v. fluids (primary therapy): rapidly, normal saline or Ringer's
solution
If poor response: consider a vasopressor, e.g. dopamine (Call
Code).

Vagal reaction (hypotension and bradycardia):
Oxygen (6–10 L/min)
Elevate patient's legs
i.v. fluids: rapidly, normal saline or Ringer's solution
Atropine: 0.6–1.0 mg i.v., repeat every 3–5 min, as needed, to 3 mg
total (adults)
(paediatric: 0.02 mg/kg i.v.; max. 0.6 mg dose; may
repeat to 2 mg total)

Anaphylaxis-like reaction (generalized):
Oxygen (6–10 L/min)
Suction, as needed
Elevate patient's legs if hypotensive
i.v. fluids: normal saline; Ringer's solution

Epinephrine
Subcut: 1:1000, 0.1–0.2 ml (0.1–0.2 mg)
(paediatric: 0.1–0.2 ml subcut)

i.v.: 1:10,000, 1 ml (0.1 mg), slowly (e.g. incrementally over 2–
5 min)
(paediatric: 0.01 mg/kg, i.v.)
* Limit epinephrine for patients taking non-cardioselective β-
adrenergic blocking drugs
* Alternative β-agonist therapy: isoproterenol 1:5000 solution
for injection (0.2 mg/ml), i.v. 0.5–1.0 ml diluted to 10 ml
with normal saline; 1 ml (20 μg) increments

Antihistamines:
H_1 blocker:
e.g. diphenhydramine 25–50 mg, i.v. (caution: may exacerbate
or cause hypotension)
H_2 blocker:
e.g. cimetidine injectable 300 mg, diluted to 20 ml, slowly i.v.
(paediatric: 5–10 mg/kg, diluted, slowly)
. . . *or* . . .
ranitidine injectable 50 mg, diluted to 20 ml, slowly i.v.
(paediatric: use not established)

β_2-Agonist metered dose inhaler (for persistent bronchospasm):
(2 or 3 inhalations)
e.g. metaproterenol, or terbutaline, or albuterol

Corticosteroids:
e.g. hydrocortisone 0.5–1.0 g i.v.
. . . *or* . . .
e.g. methylprednisolone 500 mg i.v. over 30 s, or 2000 mg over
30 min

Angina:
Oxygen 6–10 L/min
i.v. fluids: slowly, TKO
Nitroglycerine: 0.4 mg, sublingually, may repeat every 15 min
Morphine: 2 mg, i.v.

Hypertension:
Oxygen: 6–10 L/min
i.v. fluids: slowly
Nifedipine: perforate end of capsule (i.e. with a needle) and drip
liquid sublingually
If secondary to phaeochromocytoma: phentolamine 5 mg, i.v.
slowly

Seizures:
Protect the patient
Airway: suction, as needed, mouth airway for tongue control
Oxygen: 6–10 L/min
If caused by hypotension ± bradycardia, treat accordingly (per
protocols for those entities)
Uncontrolled: consider diazepam, 5 mg, i.v.

by inhaled bronchodilators. Subcutaneous administration of 0.1 mg (0.1 ml of 1:1000 solution) is usually effective treatment as long as there is adequate peripheral circulation. However, intravenous epinephrine, 0.1 mg (1 ml of 1:10 000 solution) given slowly over 2–5 minutes, should be utilized whenever there is rapid progression of symptoms or when hypotension and possible inadequate subcutaneous absorption may occur [36,66].

Hypotension

Profound hypotension may occur without respiratory symptoms. Normal sinus rhythm or tachycardia differentiates this reaction from the so-called vagal reaction (hypotension plus sinus bradycardia). In patients who have been taking β-adrenergic-blocking medications (e.g. propranolol), compensatory tachycardia may not occur. Isolated hypotension is best treated initially by rapid i.v. fluid replacement, rather than vasopressor drugs [70]. Elevation of the patient's legs is very important, in that it returns about 700 ml of blood to the central circulation and is preferable to placing the patient in the Trendelenburg position [71].

Vagal reaction

The vagal reaction is characterized by sinus bradycardia and hypotension. Although its exact cause is unknown, the vagal reaction seems to be elicited or accentuated by anxiety. Proper recognition of this reaction and its bradycardia is absolutely vital to initiating the appropriate therapy of increasing intravascular fluid volume and reversing bradycardia. The vasodilatation and expanded vascular space are treated by elevation of the patient's legs and rapid infusion of intravenous fluids; the bradycardia is treated by intravenous administration of atropine to block vagal stimulation of the cardiac conduction system. Since low doses of atropine can be detrimental in treating bradycardia associated with contrast media-induced vagal reactions [48,72–74], larger doses (0.6–1.0 mg) are indicated; my recommendation is 1.0 mg, repeated, as needed, every 3 to 5 minutes, to a total dose of 3 mg in adults [66].

Anaphylaxis-like reactions

Acute, rapidly progressing, generalized systemic reactions, characterized by pruritis, urticaria, angioedema, respiratory distress and profound hypotension, require prompt treatment, including maintenance of airway, administration of oxygen and adrenergic medications, and rapid infusion of intravenous fluids. Epinephrine is the drug of choice for these major, generalized reactions and should be administered intravenously for rapid effect and to avoid suboptimal absorption from subcutaneous tissues [36,66]. A low dose, 1.0 ml (0.1 mg) of 1:10 000 solution, is given at a relatively slow rate (over 2–5 minutes) and can be titrated to effect [75]. If the reaction is not responding to the initial, slowly administered, low i.v. dose, increase the rate of injection.

Intravenous epinephrine should be given with caution to elderly patients and in the presence of hypoxia, where there is increased risk of severe cardiac arrhythmias. Additionally, the amount of i.v. epinephrine should be limited in patients who are receiving non-cardioselective β-blocking medications (e.g. propranolol). Excessive unopposed α-adrenergic effects of epinephrine may result in an increased release of chemical mediators; in addition, the β-bronchodilatory effects are blunted. Conversely, patients who have been treated chronically with adrenergic medications may require higher doses of epinephrine for effective treatment.

When the use of epinephrine is inadvisable, bronchospasm can be treated with a β₂-agonist inhaler and hypotension can be treated with intravenous fluids. Isoproterenol, a pure β-agonist (both β₁ and β₂), with no α effects, can be used in patients on non-cardioselective β-blockers, to 'override' the blockade. The appropriate dosage can be titrated to effect; the 1:5000 i.v. solution is diluted to 10 ml and administered at 20 μg (1 ml) per minute. A small amount of i.v. epinephrine will also be necessary to achieve some α effect (vasoconstriction) and thereby correct laryngospasm/laryngoedema and angioedema.

If the patient's hypotension does not respond to aggressive intravenous fluid replacement, vasopressor medications such as dopamine may be considered. In addition, large doses of intravenous corticosteroids are often given empirically to treat anaphylactoid reac-

104

tions. Their value in treating an acute reaction is limited by slow onset of action [61].

Angina, hypertension or seizure

Treatment of these reactions after the administration of intravascular RCM is detailed in Table 2.

Cardiovascular collapse/cardiac arrest

If a patient is found unconscious and unresponsive and has no pulse or blood pressure, a 'Code' should be called and immediate CPR should be started [76].

SUMMARY

Proper preparation is most important, with equipment (Code cart, oxygen, suction, etc.), personnel (nurses, Code team personnel) and medications available immediately in any area where RCM are administered. Understanding the signs, symptoms, and treatment of various reactions to contrast media is a prerequisite for anyone taking the responsibility of administering RCM to patients. When confronted with a serious contrast reaction, immediately call for help, assess the pulse and blood pressure, and determine the exact nature of the patient's distress (i.e. difficulty with breathing, chest pain, light-headedness or cutaneous itching). With knowledge of the patient's vital signs and their major complaint, proper management of the patient can be accomplished through a logical progression of therapeutic steps.

Much of the material that appears in this article has been extracted from the following work: Bush WH, Swanson DP. Acute reactions to intravascular contrast media: Types, risk factors, recognition and specific treatment. AJR. 1991; 157: 1153–1161, with permission from the American Roentgen Ray Society; and Bush WH, Swanson DP. Radiocontrast. In: Virant F, ed. Systemic Reactions. Immunol Allergy Clin North Am. Philadelphia, WB Saunders Co., 1995, pp. 597–612, with permission of WB Saunders Co.

REFERENCES

1. Ansell G. Adverse reactions to contrast agents. Scope of problem. Invest Radiol. 1990; 25: 381.
2. Hartman G, Hattery R, Witten D, Williamson B. Mortality during excretory urography: Mayo Clinic experience. AJR. 1982; 139: 919–922.
3. Katayama H, Yamaguchi K, Kozuka T, Takashima T, Seez P, Matsuura K. Adverse reactions to ionic and non-ionic contrast media: a report from the Japanese Committee on the Safety of Contrast Media. Radiology. 1990; 175: 621–628.
4. Swanson D, Shetty P, Kastan D, Rollins N. Angiographic contrast media. In: Swanson D, Chilton H, Thrall J, eds. Pharmaceuticals in Medical Imaging. New York: Macmillan, 1990: 13–39.
5. Witten D, Hirisch F, Hartman G. Acute reactions to contrast media: incidence, clinical characteristics and relationship to history of hypersensitivity states. AJR. 1973; 119: 832–840.
6. Davies P, Roberts M, Roylance J. Acute reactions to urographic contrast media. Br Med J. 1975; 2: 434–437.
7. Shehadi W, Toniolo G. Adverse reactions to contrast media. Radiology. 1980; 137: 299–302.
8. Baltaoglu G, Balkanci R, Tirnaksiz B. Fatal reaction after intra-arterial injection of non-ionic contrast medium [Letter]. AJR. 1994; 162: 231.
9. Curry N, Schabel S, Reiheld C, et al. Fatal reactions to intravenous non-ionic contrast material. Radiology. 1991; 178: 361–362.
10. Westesson P, Manjione J. Reaction to non-ionic contrast medium during arthrography of the temporalmandibular joint [Letter]. AJR. 1990; 154: 1344.
11. Wolf G, Arenson R, Cross A. A prospective trial of ionic vs non-ionic contrast agents in routine clinical practice: comparison of adverse effects. AJR. 1989; 152: 939–944.
12. Almen T. The etiology of contrast medium reactions. Invest Radiol. 1994; 29 (Suppl 1): S37–S45.
13. Greenberger P. Contrast media reactions. J Allergy Clin Immunol. 1984; 74: 600.
14. Brasch R, Caldwell J. The allergic theory of radiographic contrast toxicity: demonstration of antibody activity in sera of patients suffering major radiocontrast agent reactions. Invest Radiol. 1976; 11: 347–356.
15. Carr D, Walker A. Contrast media reactions: experimental evidence against the allergy theory. Br J Radiol. 1984; 57: 469–473.
16. Lasser E, Slivka J, Lang J, et al. Complement and coagulation – causative considerations in contrast catastrophies. AJR. 1979; 132: 171–176.
17. Assem E, Bray K, Dawson P. The release of histamine from human basophils by radiological contrast agents. Br J Radiol. 1983; 56: 647–652.
18. Robertson P, Frewin D, Robertson A, Mahar L, Jonsson J. Plasma histamine levels following administration of radiographic contrast media. Br J Radiol. 1985; 58: 1047.
19. Littner M, Ulreich S, Putman C, Rosenfield A, Meadows G. Bronchospasm during excretory urography. AJR. 1981; 137: 477–481.
20. Cogan F, Norman M, Dunsky E, Hirshfeld J, Zweiman B. Histamine release and complement changes following injection of contrast media in humans. J Allergy Clin Immunol. 1979; 64: 299–303.
21. Lasser E. A coherent biochemical basis for increased reactivity to contrast material in allergic patients: a novel concept. AJR. 1987; 149: 1281–1285.

105

22. Lasser E, Lang J, Lyon S, Hamblin A, Howard M. Prekallikrein–kallikrein conversion rate as a predictor of contrast media catastrophies. Radiology. 1981; 140: 11–15.

23. Lieberman P, Siegle R. Complement activation following intravenous contrast material administration. J Allergy Clin Immunol. 1979; 64: 13–17.

24. Zipser R, Laffi G. Prostaglandins, thromboxanes, and leukotrienes in clinical medicine. West J Med. 1985; 143: 485–497.

25. Dawson P, Edgerton D. Contrast media and enzyme inhibition. I. Cholinesterase. Br J Radiol. 1983; 56: 653–656.

26. Lalli A. Contrast media reactions: Data analysis and hypothesis. Radiology. 1980; 134: 1–12.

27. Hayakawa K, Morris T, Katzberg R. Opening of the blood–brain barrier by intravenous contrast media in euvolemic and dehydrated rabbits. Acta Radiol. 1989; 30: 439–444.

28. Wilson A, Evill C, Sage M. Effects of non-ionic contrast media on the blood–brain barrier: osmolality versus chemotoxicity. Invest Radiol. 1991; 26: 1091–1094.

29. Lawrence V, Matthai W, Hartmaier S. Comparative safety of high-osmolality and low-osmolality radiographic contrast agents. Report of a multidisciplinary working group. Invest Radiol. 1992; 27: 2–28.

30. Caro J, Trindale E, McGregor M. The risks of death and of severe nonfatal reactions with high- vs low-osmolality contrast media: a meta-analysis. AJR. 1991; 156: 825–832.

31. Gerstman B. Epidemiologic critique of the report on adverse reactions to ionic and non-ionic media by the Japanese Committee on the Safety of Contrast Media. Radiology. 1991; 178: 787–790.

32. Kinnison M, Powe N, Steinberg E. Results of randomized controlled trials of low- vs high-osmolality contrast media. Radiology. 1989; 170: 381–389.

33. Palmer F. The R.A.C.R. survey of intravenous contrast media reactions: final report. Australas Radiol. 1988; 32: 426–428.

34. Siegle R. Rates of idiosyncratic reactions. Ionic versus non-ionic contrast media. Invest Radiol. 1993; 28: S95–S98.

35. Shehadi W. Death following intravascular administration of contrast media. Acta Radiol Diagn. 1985; 26: 457–461.

36. Bush W, McClennan B, Swanson D. Contrast media reactions: prediction, prevention, and treatment. Postgrad Radiol. 1993; 13: 137–147.

37. Shellock F, Hahn H, Mink J, Itskovich E. Adverse reaction to intravenous gadoteridol. Radiology. 1993; 189: 151–152.

38. Yoshikawa H. Late adverse reactions to non-ionic contrast media. Radiology. 1992; 183: 737–740.

39. McCullough M, Davies P, Richardson R. A large trial of intravenous Conray 325 and Niopam 300 to assess immediate and delayed reactions. Br J Radiol. 1989; 62: 260–265.

40. Lang D, Allpern M, Visintainer P, Smith S. Increased risk for anaphylactoid reaction from contrast media in patients on beta-adrenergic blockers or with asthma. Ann Intern Med. 1991; 115: 270–276.

41. Thrall J. Adverse reactions to contrast media: etiology, incidence, treatment, prevention. In: Swanson DP, Chilton HM, Thrall JH, eds. Pharmaceuticals in Medical Imaging. New York: Macmillan, 1990: 253-277.

42. Greenberger P, Meyers S, Kramer B. Effects of beta-adrenergic and calcium antagonists on the development of anaphylactoid reactions from radiographic contrast media during cardiac angiography. J Allergy Clin Immunol. 1987; 80: 698–702.

43. Fishman J, Aberle D, Moldawer N, Belldegrun A, Figlin R. Atypical contrast reactions associated with systemic interleukin-2 therapy. AJR. 1991; 156: 833–834.

44. Oldham R, Brogley J, Braud E. Contrast media 'recalls' interleukin-2 toxicity. [Letter to the Editor]. J Clin Oncol. 1990; 8: 942.

45. Shulman K, Thompson J, Benyunes M, Winter T, Fefer A. Adverse reactions to intravenous contrast media in patients treated with interleukin-2. J Immunother. 1993; 13: 208–212.

46. Zukiwski A, David C, Coan J, Wallace S, Gutterman J, Mavligit G. Increased incidence of hypersensitivity to iodine-containing radiographic contrast media after interleukin-2 administration. Cancer. 1990; 65: 1521–1524.

47. Sirtori C, Pask C. Re-evaluation of a biguanide, metformin: mechanism of action and tolerability. Pharmacol Res. 1994; 30: 187–228.

48. Fischer H, Doust V. An evaluation of pretesting in the problem of serious and fatal reactions to excretory urography. Radiology. 1972; 103: 497-501.

49. Witten D. Reactions to urographic contrast media. JAMA. 1975; 231: 974–977.

50. Siegle R, Halvosen R, Dillon J, Gavant M, Halperon E. The use of iohexol in patients with previous reactions to ionic contrast material. Invest Radiol. 1991; 26: 411–416.

51. Solomon R, Werner C, Mann D, D'Elia J, Silva P. Effects of saline, mannitol, and furosemide in acute decreases in renal function induced by radiocontrast agents. N Engl J Med. 1994; 331: 1416–1420.

52. Lautin E, Freeman N, Schoenfield A, et al. Radiocontrast-associated renal dysfunction: a comparison of lower-osmolality and conventional high osmolality contrast media. AJR. 1991; 157: 59–65.

53. Lautin E, Freeman N, Schoenfield A, et al. Radiocontrast-associated renal dysfunction: incidence and risk factors. AJR. 1991; 157: 49–58.

54. Shehadi W. Contrast media adverse reactions: occurrence, recurrence, and distribution patterns. Radiology. 1982; 143: 11–17.

55. Yamaguchi K, Katayama H, Takashima T, Kozuka T, Seez P, Matsuura K. Prediction of severe adverse reactions to ionic and non-ionic contrast media in Japan: evaluation of pretesting. Radiology. 1991; 178: 363–367.

56. Dunnick N, Cohan R. Cost, corticosteroids, and contrast media [commentary]. AJR. 1994; 162: 527–529.

57. Greenberger P, Patterson R. The prevention of immediate generalized reactions to radiocontrast media in high-risk patients. J Allergy Clin Immunol. 1991; 87: 867–872.

58. Kelly J, Patterson R, Lieberman P, Mathison D, Sevenson D. Radiographic contrast media studies in high-risk patients. J Allergy Clin Immunol. 1978; 62: 181–184.

106

59. Lasser E, Berry C, Mishkin M, Williamson B, Zheutlin N, Silverman J. Pretreatment with corticosteroids to prevent adverse reactions to non-ionic contrast media. AJR. 1994; 162: 523–526.

60. Lasser E, Berry C, Talner L, et al. Protective effects of corticosteroids in contrast material anaphylaxis. Invest Radiol. 1988; 23 (Suppl 1): S193–S194.

61. Lasser E, Lang J, Sovak M, Kolb W, Lyon S, Hamlin A. Steroids: theoretical and experimental basis for utilization in prevention of contrast media reactions. Radiology. 1977; 125: 1–9.

62. Greenberger P, Patterson R, Tapio C. Prophylaxis against repeated radiocontrast media reactions in 857 cases. Arch Intern Med. 1985; 145: 2197–2200.

63. Lasser E, Berry C, Talner L, et al. Pretreatment with corticosteroids to alleviate reactions to intravascular contrast media. N Engl J Med. 1987; 317: 845–849.

64. Dawson P, Sidhu P. Is there a role for corticosteroid prophylaxis in patients at increased risk of adverse reactions to intravascular contrast agents? [Review]. Clin Radiol 1993; 48: 225–226.

65. Luna-Fineman S, Healy M, Parker B. Corticosteroid pretreatment for potential contrast reactions in children with lymphoreticular cancer: a word of caution. AJR. 1990; 155: 357–358.

66. Bush W, Swanson D. Acute reactions to intravascular contrast media: types, risk factors, recognition, and specific treatment. AJR 1991; 157: 1153–1161.

67. Cohan R, Dunnick N. Treatment of reactions to radiologic contrast material. AJR. 1988; 151: 263–267.

68. McClennan B. Adverse reactions to iodinated contrast media. Recognition and response. Invest Radiol. 1994; 29 (Suppl 1): S46–S50.

69. Siegle R, Lieberman P. A review of untoward reactions to iodinated contrast material. J Urol. 1978; 119: 581–587.

70. van Sonnenberg E, Neff C, Pfister R. Life-threatening hypotensive reactions to contrast media administration: comparison of pharmacologic and fluid therapy. Radiology. 1987; 162: 15–19.

71. Smith M, Kendall B, Tomlinson S. Adverse general reactions to high doses of methylglucamine-based contrast media. Br J Radiol. 1974; 47: 566–569.

72. Brown J. Atropine, scopolamine, and antimuscarinic drugs. In: Gilman A, Rall T, Nies A, Taylor P, eds. The Pharmocological Basis of Therapeutics. New York: McGraw-Hill Publishers, 1990: 150–165.

73. Chamberlain D, Turner P, Sneddon J. Effects of atropine on heart-rate in healthy man. Lancet. 1967; 2: 12–15.

74. Stanley R, Pfister R. Bradycardia and hypotension following use of intravenous contrast media. Radiology. 1976; 121: 5–7.

75. Barach E, Nowak R, Tennyson G, Tomlanovich M. Epinephrine for treatment of anaphylactic shock. JAMA. 1984; 251: 2118–2122.

76. Emergency Cardiac Care Committee and Subcommittees, American Heart Association. Guidelines for cardiopulmonary resuscitation and emergency cardiac care III: adult advanced cardiac life support. JAMA. 1992; 268: 2199–2241.

This paper was first published in *Advances in X-Ray Contrast.* 1996;3:44–53.

UPDATE

Effectiveness in treatment of adverse reactions to contrast media depends to a great extent on the promptness with which therapy is instituted. No new drug has been discovered or added to our armamentarium. We already have an excellent drug, intravenous epinephrine, which I believe is underutilized in the treatment of contrast media reactions. Intravenous epinephrine (1:10 000 dilution) given incrementally in relatively small doses, and titrated to effect, is the best drug for treating many reactions. It is the most important drug for treating a generalized anaphylaxis-like reaction; even for the elderly patient, a very small amount of epinephrine added to a pure β-adrenergic drug such as I.V. isoproterenol provides excellent therapy for airway and vascular reactions. For the diffuse cutaneous reaction, epinephrine provides a most effective therapy, again given in a diluted fashion and with very small doses [1:10 000 dilution, 1 ml (0.1 mg) titrated slowly over 2–4 minutes].

P. Dawson and W. Clauss, (eds.), Advances in X-Ray Contrast: Collected Papers. 107–110
© *1998 Kluwer Academic Publishers.*

OVERVIEW

Contrast medium administration in spiral computerized tomography: an overview of a consensus meeting in radiodiagnosis

S Feuerbach, PhD, MD
Director of the Institute for Radiodiagnosis, University Regensburg, Germany

INTRODUCTION

Considerable variations exist in the routine use of spiral computerized tomography (spiral CT) in terms of dose, flow rate and scan delay. There are also substantial differences cited in the literature for the volume of contrast media (CM) used in spiral CT. In order to provide a consistent approach for the use of CM in spiral CT, a Consensus Conference entitled 'Contrast Medium Administration in Spiral Computerised Tomography' and sponsored by Schering AG was held in Berlin on 17–19 February 1995. The aim was to draw up guidelines for CM administration (volume, flow rates, scan delay) and choice of scanning parameters (slice thickness, table advance and calculated reconstruction interval) in spiral CT.

The conference faculty and delegation comprised a group of 21 speakers from 17 radiodiagnostic institutions in the Federal Republic of Germany, who were experts in the use of spiral CT in their chosen field. They represented current opinion in the areas of radiology, surgery and clinical methodology. Additionally, there were 26 audience participants, some of whom were eminent radiologists, together with media publishers and representatives of industry. The following overview provides a summary of the key facts to emerge from the Consensus Conference*.

METHODOLOGY

The Conference discussion focused on eight regions of the body, namely the neck, thorax, liver, pancreas, retroperitoneum, kidney, pelvis and vessels. Two further topics covered by the speakers were the study technique for combined examinations of several anatomical regions in one imaging session and the prerequisite for post-processing a CT dataset (e.g. 2- or 3-dimensional reconstructions). The following key questions were addressed by the speakers:

1. What are the criteria for optimal contrast?

2. What is the basis for these criteria?

3. What is the optimal CM dose? What iodine concentration is required? Are any safety margins built into this form of administration?

4. What flow rate and scan delay are required? Is a power injector a prerequisite?

5. What are the essential parameters for spiral CT?

CONSENSUS

Overall, a total of 119 topic-related questions were addressed by the delegation. The beliefs and judgements of each participant were obtained anonymously immediately following each presentation and discussion session. The answers obtained were collected and printed out for participants to reconsider on a second occasion. Eventually, a final consensus of all

*The material that appears in this article has been extracted from the following work: S Feuerbach, W Lorenz, K-J Klose, J Gmeinwieser, K-J Lackner, P Landwehr, E Grabbe, R Klöppel. Kontrastmittelapplikation bei der Spiralcomputer-tomographie: Statement einer Konsensus-Entwicklungs-konferenz in der Röntgendiagnostik. Fortschr Röntgenstr. 1996; 164 (2): 158–165, with permission from Georg Thieme Verlag, Stuttgart, New York.

108

participants was reached on each topic and expressed by organ, in percentage terms. The main results for each body region are summarized below:

Neck

- 92% of participants considered the criteria for obtaining optimal contrast are:
  (i) a vascular density of approximately 150 HU, and
  (ii) interstitial CM uptake.
- 78% of participants believed that the optimal contrast medium dose is 100 ml.
- 84% or more participants agreed that a flow rate of >1.5 ml is necessary, with a scan delay of <50 seconds and table feed and reconstruction interval of 5 mm/s.
- Only 27% of participants felt that economic aspects are important in their choice of CM.
- As few as 3% of participants considered that a plain CT scan is adequate for the neck region, with 59% believing that spiral CT is necessary to answer the diagnostic questions raised at this conference.

Thorax

- 76% of participants considered the criterion for obtaining optimal contrast is a high intravascular density of the mediastinal vessels.
- 58% of participants considered the optimal volume of CM to be 60 ml, whilst 32% preferred a higher dosage for differentiation of mediastinal structures.
- 87% of participants considered a caudocranial projection to be optimal for achieving recirculation with sufficient contrast of both arteries and veins above the base of the valves.
- 95% of participants believed that spiral CT is necessary to answer the diagnostic questions fully.

Liver

- The participants were unanimous (100%) in believing that the criterion for optimal contrast in parenchymal enhancement is $\geqslant 35$ HU and that the ideal arterial scan delay is $\leqslant 30$ seconds.
- The ideal venous scan delay was considered to be between 60 and 70 seconds by 97% of participants.
- 82% of participants felt that a plain CT scan is

usually adequate for the liver region, with 74% preferring a slice thickness of 8–10 mm.
- A caudocranial scan direction is thought preferable by 54% of participants.
- 89% of participants believed that the optimal volume of CM for the liver region is $\leqslant 150$ ml, with 50% considering the preferred flow rate to be $\leqslant 2.5$ ml/s.
- 95% of participants believed that spiral CT is necessary to answer the diagnostic questions fully.

Pancreas

- 82% of participants were in agreement that contrast enhancement of the pancreatic parenchyma to the extent of $\geqslant 80$ HU and of the portal vessels to the extent of $\geqslant 180$ HU are optimal for this organ.
- 95% of participants felt that a spiral CT scan is necessary to answer the diagnostic questions fully; however, 53% believed that a non-enhanced scan is usually needed for a baseline value.
- 97% of participants advocated a slice thickness of $\leqslant 5$ mm.
- A scan delay of 70 seconds is felt necessary to assess the portal system by 86% of participants.
- 87% of participants believed that the liver must be completely imaged in pancreatic carcinoma and acute pancreatitis.
- 38% of participants agreed that the optimal CM volume is $\leqslant 130$ ml; 62% preferred either a smaller or higher volume.

Retroperitoneum

- Only 13% of participants felt that an unenhanced CT scan is adequate for this region.
- More than 92% of participants believed that a CM dose of 100 ml, with a flow rate of 2 ml/s and scan delay of 50 seconds, are expedient, and that a caudocranial approach is appropriate.
- A slice thickness of 8 mm, with table advance of 12 mm/s, is favoured by over 80% of participants.
- 71% of participants felt that a spiral CT scan is necessary to answer the diagnostic questions fully.

Pelvis

- The majority of participants believed that the pelvis should be examined simultaneously with the abdomen and that the examination technique is defined primarily by the abdominal organs.
- As a result, discussion of the pelvic region as a separate entity was curtailed.

Kidney

- 81% of participants agreed that the criteria for an optimal examination are initial corticomedullary discrimination and subsequent imaging of the kidneys in the parenchymal phase.
- 97% of participants considered an unenhanced scan to be mandatory for the kidney region (90%, flow rate 2 ml/s; 95%, optimum CM dose about 100 ml and slice thickness ≤ 5 mm).
- There was unanimous agreement that subsequent dual-phase examination should be performed.
- 94% of participants felt that spiral CT is necessary to answer the diagnostic questions fully.
- In two-phase studies, the first scan delay is ≤ 30 seconds (86%) and the second scan delay ≤ 5 minutes (100%).

Blood vessels

- There was unanimous agreement that the criterion for optimal contrast is an intravascular density of ≤ 200 HU.
- There was excellent agreement amongst participants on the following aspects of vessel scanning: CM dose 100–150 ml; scan delay defined by the bolus; slice thickness (small) 2–3 mm, (large) 3–5 mm; flow rate ≥ 3 ml/s. The scan delay should be defined by bolus tracking (83%).
- 89% of participants felt that spiral CT is necessary to answer the diagnostic questions fully.
- 49% of participants felt that economic aspects played a role in their choice of CM dose for the vessel region.

2-D and 3-D post-processing, bolus administration

- Only 19% of participants felt that 2-D and 3-D post-processing are necessary for the demonstration of tumours; however, 93% of participants considered the further development of these techniques to be essential, preferably employing a separate work station for their use.
- 71% of participants considered the incorporation of automatic bolus administration in spiral CT scanners to be expedient.
- For CT angiography, as many as 83% of participants advocated the definition of delay by automatic bolus administration; economic considerations do not allow for routine 2-D and 3-D reconstructions.

Combined examination

- 69% of participants felt that combined examination is expedient for tumour screening.
- At least 95% of participants agreed that the technique was expedient in polytrauma and for the staging of lymphomas and the demonstration of aortic aneurysm.

CONCLUSION

There was a consensus (range 89–97%) among participants that spiral CT was the ideal modality for imaging of the anatomical regions of thorax, liver, pancreas, retroperitoneum and kidney. 59% of participants felt that this was also the case for the neck region.

The conference unanimously agreed upon the following recommendations for the organs discussed, in terms of CM administration (regardless of volume, scan delay and flow rate):

- Only non-ionic CM should be used, to avoid any possibility of motion artefacts caused by vomiting, nausea or a sensation of warmth, which may prevent calculation of a spiral CT dataset.
- Power CM injectors should be used exclusively, in order to maintain the necessary constant high flow rate and facilitate a homogeneous bolus.
- ≥ 18-gauge needles should be used in the cubital vein.

110

Further important conclusions from the conference are summarized below:

- Unenhanced scanning techniques are unacceptable for the neck, thorax, liver, pancreas, kidney and non-organ-bound retroperitoneum regions, for qualitative reasons.
- CM volumes agreed by consensus represent guidelines regarding dosage, with a recommendation to use the lower normal values wherever possible. The use of CM volumes below the lower normal values is not recommended, for qualitative reasons.
- The acceptance of these guidelines should improve the quality of CT examinations and reduce the number of repeat examinations.

- The documented advantages of spiral CT must, in the long term, lead to the replacement of obsolete scanners.

In terms of cost, it is imperative that optimal techniques for CM administration should always be researched and developed without initial emphasis upon economic aspects. However, the economic implications of choice of CM play a minor role in most anatomical regions, and must be taken into consideration during clinical practice.

This paper was first published in *Advances in X-Ray Contrast*. 1996;3:66–69.

P. Dawson and W. Clauss, (eds.), Advances in X-Ray Contrast: Collected Papers. 111–114.
© 1998 Kluwer Academic Publishers.

Index

N-acetyl-β-glucoasminidase 14, 15, 16
β-adrenergic blockers 98–9
  aetiology 97–8
  mortality rates 98
  prevalence 98
  prevention 99–104
    chemotoxic reactions 99–100
  risks 98–9
    allergy 98
    asthma 98
    medications 98–9
    prior reaction 99
    total dose 99
  treatment for specific reactions 101–4
    anaphylaxis-like reactions 103–4
    angina 104
    bronchospasm 101
    cardiac arrest 104
    cardiosvascular collapse 102
    cutaneous reactions 101
    hypertension 104
    hypotension 103
    nausea 101
    seizures 104
    vagal reaction 103
    vomiting 101
    typical 101
β-agonists 101, 103
alanine aminopeptidase 14
albuterol 101
alcoholism, chronic 57
alkaline phosphatase 14, 16 (fig.)
allergy 98
Almén, T. 93–4
angiocardiography 41
angiography 41
angio-oedema 30, 31
anticoagulants 21
antidiuretic hormone 5
antihistamines 100
aortic aneurysm 74
aorto-coronary bypass graft 72
arm pain, delayed 31
arteriography 57
arterioportography 57, 61
asthma 31, 98
atropine 103

back pain 31

barium 38–9
benadryl 54
benzoic acid derivatives 92–4
blood clots 20
bone fractures 66
bronchodilators 101
brush border-related proteins 14

calcium 47, 50–1, 52
  antagonists 98
Cameron, D. 90
carcinoid 59–60
cardiac arrhythmia 21, 49
cardiac catheterization 50
cardiac effects of contrast agents 46–51
  clinical 49–50
    arrhythmias 21, 49
    thrombosis 49
    ventricular fibrillation 49
  cost 50
  experimental 46–9
    coronary arteries 46, 47 (fig.)
    left ventricle 48
  utilization 50
catheter
  flushing technique 25–6
  thrombogenicity 20, 21–2
chelate calcium 47
chemotoxic reactions 35
children 33, 34–43
  accidental overdose 35
  adverse effects of contrast media 34–6
  anaphylactoid-type reactions 36
  angiocardiography 41
  angiography 41
  body computed tomography 42
  bowel perforation 40
  bronchography 43
  central nervous system 42–3
  choice of contrast medium 34
  diabetic 35
  extravasations 35, 38 (fig.)
  gasless abdomen 40
  gastrointestinal tract 38–9
    barium 39
    choice of contrast agent 38–9
  intussception reduction 40
  iodinated contrast materials 86
  meconium ileus 40

112

children *(continued)*
  myelography 35, 42–3
  necrotizing enterocolitis 39, 40
  neonates 35, 40
  osmolality of intravenous contrast agents 35 (table)
  pneumoperitoneum 40
  pressure (power) injection 35
  pulmonary oedema 35
  renal dysfunction 35
  treatment of adverse reactions 36, 37 (table)
  tumour lysis prone 36
  voiding cystourethrography 36
    cystitis due to 36
  water-soluble contrast material 39–40
cimetidine 54
clinical angiography 90–1
  thromboembolic phenomena 20–7
Conray 93
contrast agents 22–3
  nephrotoxicity associated 3
contrast enhancement in CT of liver, pancreas, spleen
        57–62
  complications 61
  contrast medium 57–8
coronary angiography 22, 50
coronary occlusion 20, 21
corticosteroids 100
creatinine 4, 8, 14
Cysto-conray 36

delayed reactions to intravenous injections 29–33
dermo-epidermal necrosis 30
diabetes mellitus 50
diatrizoate 36
diatrizoic acid 81
digital angiography 54
digital fluoroscopy 34
digital systems 54
dionosil 43
diphenhydramine 84, 100

EDTA 47
electron beam computed tomography 67–80
  advantages 69
  aortic aneurysm diagnosis 74–5
  application protocols 70–3
  cardiogreen dye method 71
  circulation time 72
  contrast protocol 75
  heart imaging 76–8
    contrast media application 78
    endocardium 78
    myocardial function 77
    myocardial mass 78–9

    myocardial perfusion 77
    myocardial shape 76–7
    pericardium 76
  iodine content 70
  magnesium sulphate method 71
  method of application 71
  operating modes 67–9
    cine mode 68
    continuous volume mode 69
    flow mode 68
    multi-slice method 67–8
    single-slice method 67
  pulmonary embolism diagnosis 79–80
  scanner 67, 68 (fig.)
  scanning protocols 71–3
    abdomen 72
    head and neck 73
    heart (flow study) 71–2
    heart (volume study) 72
    kidneys 73
    pelvis 73
    retroperitoneum 73
    thoracic study 71
  toxicity 70
  types of contrast agents 70
embolization 20, 54
  mixtures 54
enzymuria 17–18
ephedrine sulphate 100
epinephrine 101–2, 103, 106

femoral angiography 12
'flu-like' illness 29, 30, 31
flushing agents 22
Forssmann, W. 92

gadolinium chelates 94
Goodspeed, A.W. 89
Gastrografin 40
  complications 40
gastrointestinal disturbances 31
glomerular filtration rate 4, 11, 18
  contrast media effect 12–14
guidewire thrombogenicity 22

haemoglobin SC 85
haemoglobin SS 85
headache 31
heart failure 29
heparin 21
heparinization 22, 6
hepatic artery 60–1
  accessory 60–1
hepatocellular carcinoma 52

hepatoma 59–60
high-risk patients 8
hydralazine 33
Hypaque 36, 40, 93
hyperosmolality 81
hypertension 50
hyperthyrotic patients 70
hyperuricaemia 85

incompatibilities 54
indomethacin 18
inorganic iodide 55
insulinoms 59
interleukin-2 99
interventional radiology 52–5
  abnormal treatment of contrast agents 54–5
    extreme conditions 54
    lasers 55
    mixing with other agents 54
  contrast agents, chemistry/pharmacology 53
  high-dose toxicity 53
  idiosyncratic reactions 53
  theoretical considerations 53–4
intraidol 58
intravascular contrast agents, first 100 years 89–95
intravenous urography 91–2
intussusception reduction 40–1
iodinated contrast agents 81–6
  Bowman Gray Medical Center 86 (table)
  contrast material doses in neuroradiology 86
  nephrotic potential 3–9
  paediatric patients 86
iodide mumps 29, 30
iodides 29
  vegetating 30
iodinated contrast agents 81–6
  adverse contrast medial reactions 82–3
    chemotoxic effects 82
    idiosyncratic reactions 83
    neurotoxicity 82–3
    special circumstances 85–6
  adverse reactions rate 83–4
    non-ionic vs. ionic contrast materials 83–4
    treatment 84–5
  nephrotoxicity 84
  prophylactic treatment 84
iodine 15, 90
iodism 31
iodixanol 17, 88, 94
iohexol 17, 33, 81, 94
iopamidol 23, 78 (fig.), 81, 94
iopentol 13–14, 17
iopromide 81
iothalamate 18, 36

iothalamic acid 81
iotrolan 46, 94
ioversol 81
ioxaglate 23, 82, 94
isotypes clearance 12

Jennings, W.N. 89

kidney
  contrast handling 4–5
  enzyme excretion 15–17
  glomerular injury 6
  pathophysiology of adverse contrast effects 5–7
  perfusion 5
  tubular cell injury 7
Kollargol 91

lasers 55
Lichtenberg, A. von 91
Lipiodol 57, 58, 61
  acute cholecystitis following 61
liver 59–61
  arteriography 60–1
  arterioportography 61
  cirrhosis 57
  contrast enhancement 57
  delayed hepatic computed tomography 60
  dynamic hepatic computed tomography 59–60
  electron beam computed tomography 73
  focal lesions 73
  helical computed tomography 60
  metastatic disease 52
  spiral computed tomography 51–2
low osmolality agents 3, 8, 47
los osmolar x-ray contrast media 11, 13, 14
  children 39–40

mannitol 15
maxillofacial injuries 66
meconium ileus 40
metaproterenol 101
metformin 99
methylprednisolone 100
metrizamide 6, 40, 43, 93–4
Moniz, E. 90–1
myeloma 70
  multiple 85
myocardial infarcts 21
myxomas 78

nephropathy contrast media induced 12–13
nephrotoxicity 3–9, 11–18
  animals models 7–8
  contrast agent induced 3–4

nephrotoxicity (*continued*)
   LOCM/HOCM compared 14
niopam 30
non-ionic agents, allergic reaction 33
non-ionic dimers 28, 52

obstructive nephropathy 7
osmolality 81
osmotic nephrosis 17

pancreas 58–9
  carcinoma 57
  helical computed tomography 59
  islet cell tumour 59
  vascular neoplasms 59
pancreatic mumps 30
papaverine 54
parotitis 30, 31
phaeochromocytoma 85, 86
pharmacokinetics 11–12
phenformin 99
phlebitis, delayed 29
platelets 25, 28
polypropylene 24
post-mortem angiographic contrast media 90
prednisone 84, 100
protamine 54

red cell aggregation 23
renal cell carcinoma 59–60, 73
renal failure, acute 31
renal function impairment 3
renal tubular function 14–18
risk factors 4, 13
Roentgen, W.C. 89–90

sclerotherapy 54
Selectan Neutral 91
serum-sickness-like syndrome 31
'shuttle' scanning 64
sickle cell anaemia 85
skin eruptions 31
sodium cations 52
sodium citrate 47, 48 (fig.), 52
sodium EDTA 48 (fig.), 52
sodium meglumine diatrizoate 47 (fig.)
sodium meglumine ioxaglate 52–3

sodium monoiohippurate 92
spinal osseous pathology 65
spiral computed tomography 63–6, 107–10
  aortic examination 64
  blood vessels 109
  combined examination 109
  contrast media 65
  2-D/3-D post-processing 109
  kidney 109
  liver 51–2, 108
  methodology 107
  neck 108
  pancrease 108
  pelvis 109
  retroperitoneum 108
  thorax 108
  trauma 65–6
spleen 58
Stevens–Johnson syndrome 30
Swick, M. 91–2
syringes, thrombogenicity 20, 23, 24 (fig.)
systemic lupus erythematosis 30

terbutaline 101
thorium arteriography 91
thrombin time 25
thrombocytopenia, acute 30
thromboembolic phenomena 20–7, 56
thrombus formation 24–5
thyroid disease 85
thyrotoxicosis 86
toxic epidermal necrolysis 30
trauma 65

urate nephropathy 88
urinary enzymes 14
Urographin 93
urography 3
urinary β_2-microglobulin 14
Uroselectan 81, 91
urothelium 33
urticaria 33, 101

vasculitis 30, 33
  necrotizing 30
vasopressin 6